D1020092

On Methuselah's Trail

On Methuselah's Trail

LIVING FOSSILS
and the
GREAT EXTINCTIONS

Peter Douglas Ward

W. H. FREEMAN AND COMPANY
New York

Illustrations by Linda Krause, unless noted otherwise.

Library of Congress Cataloging-in-Publication Data

Ward, Peter Douglas, 1949–
On Methuselah's trail : living fossils and the great extinctions / by Peter Douglas Ward.
p. cm.
Includes index.
ISBN 0-7167-2203-8
1. Living fossils. 2. Extinction (Biology) I. Title.
QL88.5.W37 1991
591.3'8—dc20 91-17071
CIP

Printed in the United States of America

1 2 3 4 5 6 7 8 9 VB 9 9 8 7 6 5 4 3 2 1

For Nicholas, Angela, and Joe

Contents

Foreword

by Steven M. Stanley

Perhaps we should not be surprised that Charles Darwin coined the phrase "living fossils," since he cleverly addressed so many of the curious features of evolution in constructing a defense for his revolutionary ideas. Living fossils have been variously defined, but by any definition they are sole survivors—small groups of animals or plants that are the only living representatives of geologically ancient categories of life. Living fossils share another remarkable attribute: they seem frozen in time, closely resembling relatives that lived tens or hundreds of millions of years ago. Two French authors colorfully described such organisms as having "stopped participating in the great adventure of life."

Peter Ward offers *us* an adventure. It is one of vicarious discovery—discovery of living animals and plants once thought to have vanished from the earth eons ago; discovery of the biological traits of strange, long-departed, and imperfectly fossilized species; and discovery of the secrets of survival in the great game of nature in which losing means annihilation. The living fossils that have roles in Ward's story range from coelacanths of the ocean deeps to horseshoe crabs at the edge of the sea to towering redwoods on the land.

The occasional serendipitous discovery of living fossils sparks our imagination. From time to time every paleontologist harbors a secret fantasy in which a wondrous new species of a biological group thought to be extinct turns up in a dense tropical forest or deep-sea trench. Discoveries of living fossils impel the general public even further, toward science fiction. They add a measure of credence to claimed sightings of the Loch Ness monster and Big Foot. Might "Nessie" be a plesiosaur from the Age of Dinosaurs? Probably not, since the Loch Ness basin lay beneath thick glaciers during the Ice

Age. Might Big Foot or the Abominable Snowman be a mountaineering Neanderthal? Not likely, since none of Neanderthal's distinctive stone tools litter the Cascade Ranges or Himalayas.

In the realm of real science, a living fossil is a kind of time machine that allows us to glimpse part of a lost biological world. Twenty years ago, I made a personal voyage of discovery to Australia, the only region of the globe now inhabited by the *Trigoniidae*, a family of marine bivalve mollusks that resemble modern cockles but which flourished only prior to the Age of Mammals. These clams were once thought to have gone the way of the dinosaurs: they were known only from their fossilized shells. *Neotrigonia*, the lone surviving genus, was not discovered until 1802. It includes just five species, each of which occupies a segment of the ring of shallow sea floors that encircle Australia. As the sole survivor, *Neotrigonia* seemed the only vehicle by which I might explain the bizarre shell shapes of its extinct relatives—shapes that include unusual profiles, huge interlocking hinged teeth, and unique ridges or rows of knobs on the outer surface.

I discovered that, by virtue of a highly muscular foot, *Neotrigonia* can burrow into sand more rapidly than many other clams. It can even jump when disturbed. Unlike more sluggish clams, it can inhabit shifting sands where waves or currents occasionally dislodge animals of its type, which live shallowly buried. Reburrowing to the safety of the sand is no problem for *Neotrigonia*. Its agility often keeps it one step ahead of a nearby predator. In the muscular foot and associated behavior of this living fossil, I had the key to understanding the extinct members of its strange clan. From Texas to England and France, I had chipped their fossilized shells out of coarse-grained rocks that had once been shifting sands. Obviously, natural selection produced the large foot to accelerate the burrowing process. By observing *Neotrigonia* I could now see that the large hinged teeth, present in this living form as well as in extinct species, served to keep the shell halves aligned when the foot emerged. On the negative side, the bulky hinged teeth prevented members of this family from evolving the "beak-forward" shell shape that assists most other kinds of clams in the mechanics of burrowing. Saddled with a poorly adapted general shape, members of the *Trigoniidae* evolved compensatory structures—a variety of unusual ridges and knobs, with each configuration providing an adaptive solution for a particular species.

Back at home, I built a machine that forced robots, cast from hundred-million-year-old fossils, to "burrow" into sand on the floors of laboratory aquaria. Robots filed smooth made slower progress than natural ones, whose ribs and knobs gripped the sand as the artificial shells rocked back and forth. My robots mimicked real animals burrowing into sea floors that bordered the continents where dinosaurs roamed. Trigoniids finally made sense. Their several curious features were interrelated, having evolved as a coadapted complex that could be traced back to the incipient evolution of the muscular foot, seen today only in *Neotrigonia*.

Peter Ward tells how a portion of his research followed a similar path. The targets of his inquiry were the ammonites, relatives of the living nautilus. The ammonites died out with the dinosaurs, along with nearly all of my trigoniid friends. Just as there is only a handful of living species of *Neotrigonia*, there are but a few extant species of *Nautilus*, which also happen to be confined to the eastern Pacific and Indian oceans. Like its relatives the squids and octopuses, nautilus is a jet-propelled predator, catching prey in its tentacles and biting off chunks of flesh with a parrotlike beak. Ward has studied how nautilus remains buoyant, despite its dense shell, by pumping liquid from each new chamber and replacing it with gas. The extinct ammonites were similar to nautilus, as well as to extinct nautiloids, in both form and physiology. In fact, the ammonites evolved from the nautiloids, only to be outlived by them. Ward suggests that the nautiloids escaped extinction because their offspring lived safely at depths in the sea, whereas ammonite progeny, which floated as plankton, died in the collapse of the planktonic food web that seems to have accompanied the disappearance of the dinosaurs.

Here, Ward's story goes beyond mine. He has a viable hypothesis for the preferential extinction of the ammonites. I cannot easily explain the near extinction of the agile *Trigoniidae*, except to note that many were restricted to the tropics, where especially heavy losses of life occurred when the Age of Dinosaurs came to a sudden end. Ward also recounts his fieldwork in Spain, where he tracked the last of the ammonites upward through layers of rock to find their records terminate abruptly. Their end came close to the level where a high concentration of iridium points to the catastrophic impact on earth of a large meteorite or comet.

Ward presents us with much more than a roster of living fossils. In fact, he offers a voyage through geologic time. He associates the Methuselahs, as he calls them, with their ancient relatives, and he re-creates the worlds in which their forebears flourished and then died. When paleontologists study living "fossils" in order to inject life into real fossils, they are adhering to the principle of actualism, or the investigation of earth history in light of entities and processes observed in the modern world. Ward illustrates how modern paleontologists work within this framework, studying how ecosystems have undergone drastic changes—sometimes slowly, even on a geological scale of time, and sometimes abruptly.

Ranging beyond tales of evolution and extinction, Ward offers a celebration of fieldwork. His nostalgic personal accounts move from the ammonite-bearing strata of California and Spain to the more ancient fossil-packed limestones of the Cincinnati region to the blue-water haunts of nautilus that fringe the tropical paradise of Vanuatu. In the process, Ward depicts the romance of paleontology as powerfully as he conveys its unique intellectual contributions.

1

Introduction
The Phenomenon of Living Fossils

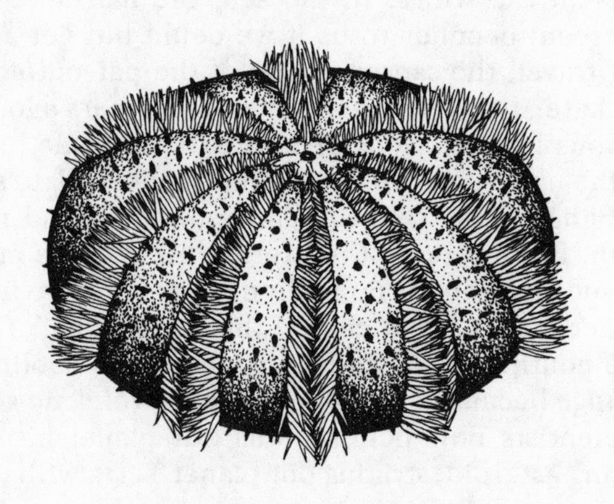

The Château Bellecq

I am driving north along the N112 in southern France in late-May sunshine, moving toward the next destination printed in the Michelin road guide to the geology of the western Pyrenees. My obvious pleasure at this fine day is slightly tempered by the madmen piled up on my rear bumper, demanding an additional 50 kilometers an hour to my already breakneck speed. They pass by, one by one, angrily tooting their horns. But I cannot be vexed by these terrible drivers; the months of teaching in gray Seattle have been replaced by the greenness of France, and I am engaged in a fascinating quest.

I am on the track of exposures of sedimentary strata deposited at the end of the Mesozoic Era, hoping to learn something about the causes and consequences of the extinctions that so changed our world. According to my trusty guide, there will be suitable rock exposures of about this age fifteen kilometers ahead, in a small town called Bellecq. I find my exit and turn onto a narrow tree-lined avenue. The twisting route gives way to fields, then a small village, and finally the river I seek. I stop my rented Fiesta to stretch and take another look at my guide. My several days of exploring outcrops in the Pyrenees have not been particularly productive from a scientific point of view. I want to find additional stratigraphic sections that expose the boundary between the Mesozoic strata and the overlying Cenozoic strata.

The boundary between these two major units of geological time has been the subject of intense debate for over a decade. The Mesozoic world was populated by dinosaurs on land, with our direct ancestors, tiny ratlike creatures called stem mammals, living fearfully in the wings. In the sea, the marine communities would also seem peculiar to us if we could but get a glimpse of them (time travel, the sacred dream of the paleontologist!). The seas of the Late Mesozoic, about 70 million years ago, were filled with wondrous but now long-dead creatures: coiled ammonites, cousins of the nautilus, with ornamented saucer-plate shells; huge flat clams lying on the seabed; shoals of squid and now-extinct forms of fish. But mystery of mysteries, all of these marvels died out rather suddenly about 66 million years ago, leaving behind a much emptied world. What caused this great dying? Did the seas flow off the continents, sparking a rapid global cooling? Did the climate change because of suddenly intense volcanic activity? Or, as many scientists now believe, was our planet hit by a huge, earth-crossing asteroid, striking our planet Terra with cataclysmic force sufficient to kill 50 percent of the earth's species on both land and sea? The answer to the riddle of the great extinction lies buried in the rock record of that ancient time. But rocks that preserve this small interval of time of so long ago are rare; a few sites in North America, one in Antarctica, and several others scattered across Western Europe provide our best evidence. My goal on this trip is to search out other places where the passage from the Mesozoic to the succeeding Cenozoic Era is exposed in sedimentary rock. And I hope that such a place will be found in this sleepy French village called Bellecq.

My map suggests that if 65-million-year-old rocks exist in this region, they will be around the next turn of the river. I return to my car and wind my way through the narrow streets of Bellecq. Life moves lazily in this medieval village, the slow roll of centuries visible in the stone buildings and cobbled roads. I turn past a high stone wall, following signs toward some château, thinking of outcrops, not castles. As I come around a final bend the village drops away. A blinding whiteness sears my eyes after the shadow of the high-walled village road. Perched above a broad turn in the river sits an immense stone castle, constructed of brilliant white lime-

The Château Bellecq.

stone dazzling in the lazy afternoon sunshine. It appears very old, its high tower crumbling; here and there green vines snake through holes in the massive walls. But even among the signs of decay the shear whiteness of the edifice speaks of renewal. I approach this immense monument, now on foot, cowed into leaving behind the tools of my trade, the cold steel of hammer and chisel which are my access back into time.

The château, built in the thirteenth century, is closed up, its grounds deserted. Although I have just stepped from twentieth-century France I am curiously isolated next to the gleaming walls, accompanied only by the songs of the robins perched around me. I walk alongside the walls of this great castle until I finally reach the river. The banks of the slowly moving river show dark-brown strata along their edges; I descend to these mossy shales and gently dislodge several pieces. My usual modus operandi is lustily to blast away with a rock hammer, but that seems inappropriate with the huge white walls looming over me. I look closely at the shales and find the signs I seek: small pieces of ammonite fossils in the highest of the shales, undeniable evidence of the Mesozoic Era. And it is from these squat dark rocks of the river that the château rises, white roots of rock like immense bony toes rushing upward into the walls and turrets standing over me, the white bricks composed of pure limestone of the earliest Tertiary age, the first rocks deposited after the cataclysm that ended the Mesozoic. I turn my attention from the dark shales of the river to the walls of the château and the limestone of which they are made. I have seen this limestone before on the coasts of France and Spain, at magical places called Zumaya, Sopelana, Hendaye, and Bidart, places where the last Cretaceous and the earliest Tertiary systems of sedimentary rock lie in visible contact. But in those places the transition from the middle age to the new age of life is seen along sea cliffs, and is manifested simply as a layer of white Cenozoic limestone sitting atop the last dark Mesozoic shale. At Bellecq the hand of man has taken this contact layer and shaped it. The ancient inhabitants of Bellecq disdained the dark marls of Cretaceous age as the framework for their great monument, instead favoring the white limestone lying above. I look more closely at the limestone, and the bricks come alive. Within them I find round objects, slightly smaller than golf balls, with a regularity of shape and pattern that comes only from life. I gently remove one of the round objects from the side of the castle. It is an echinoid, a sea urchin, 65 million years old. It looks much like the urchins I have seen in the tropical waters of the western Pacific, and similar as well to the urchins of the underlying Cretaceous rock. Of all of the life to be found in the underlying Cretaceous shales, from ammonites to giant clams to single-celled plankton, only this round echinoid species survived the cataclysm in this region of the world. And like the lowest foundations of the castle, rising up from the dark mossy shales of the river, this small morsel of life is an early piece of the great tree of life that grew out of the ashes of the end of the Mesozoic. I like to think of it as a Methuselah.

One of my favorite quotes, always good for a laugh in my paleontology classes, comes from Columbia University paleontologist Norman Newell, who once summarized mass extinctions in a way it's hard to argue with: "Death has a higher probability than immortality." Newell was referring, however, not to the life span of an individual organism but to the life span of a species.

The modern concept of species was developed in the seventeenth century by the English naturalist John Ray, who considered that organisms should be grouped on the basis of common descent, similarity of features, and constancy of those features through reproduction. This definition of "species" is not very dissimilar to the one advanced in 1942 by the great biologist Ernst Mayr: "Species are groups of actually or potentially interbreeding populations [of organisms] which are reproductively isolated from other such groups." The key point is that species are distinguished by the ability to breed successfully from generation to generation.

This definition, although admirable for currently living organisms, is obviously useless to anyone who is studying fossils. Though the fossil record yields many insights into the mode of life of extinct organisms, it simply doesn't tell us the juicy details of who was sleeping with whom back in the Mesozoic. Nevertheless, paleontologists refer to their fossils as belonging to species, and in doing so are implying that the various individuals they place within their species could, when alive, breed successfully. But the truth is that fossils are grouped in species solely because of morphological similarity. In other words, they are similar in form and structure.

In many respects the life of a species is like the life of an individual: a species is "born" from an immediate ancestor, its "mother"; and finally it dies through the process of extinction when its last individual member dies.

Extinction. The concept seems so much more horrifying than the simple death of a single organism. Extinction eliminates an entire gene pool, the sum of genetic information that maintains a species. Either gradually or quickly, the individual members of the various populations of the species die off, their numbers dwindling as reproductive rates lag behind death rates. The gene pool contracts as the population diminishes. Perhaps the reduction in numbers of individuals is a long-term process, taking millions of years, with short-term increases in numbers masking longer-term decline. Or perhaps the reduction is virtually instantaneous (at least in geological time), with all members of the species dying in sudden fire or chemical warfare.

Extinction is the fate of all species. And it is not an abstraction. At some moment in time there was but a single dinosaur left on earth, or a single ammonite, swimming in the sea, just as there will soon be a time when but a single California condor will be left on earth, vainly seeking a mate. And when that last individual dies, the unique genetic information that makes up its kind will disappear. Our species will certainly not be exempt. Even if we break free of our solar system and populate a million other worlds, there will be a time, probably not long from now (in geological time), when a single human is left, the last of our contentious heritage, the last of our species, the end of our gene pool.

Death comes to an individual for a variety of reasons. For the vast majority of organisms on earth, death comes at the hands or fangs or action of some predatory creature or microbe, for very few organisms die of old age. Indeed, many creatures could probably be relatively immortal if they were given the chance; sea anemones, for instance, seem to show no signs of aging if they are kept in a laboratory environment shielded from predators and changing environmental conditions. But such conditions are seldom found in nature.

And what of species? Can a species die of old age? Has any species died out because it has grown old, going peacefully in its sleep, as it were? Death comes to a species, as to an individual organism, for any of a variety of reasons: the evaporation of a lake, the change in temperature of a seaway, the bulldozing of a rain forest, the loss of a necessary food supply, the introduction of a new predator; the list of possibilities is long. Old age, however, does not appear to be one of them. Earlier in this century a popular school of evolutionary thought suggested that species could indeed "die" of old age, and that the last generations of species of long-lived families or orders of creatures manifested bizarre signs of their antiquity. This concept of "racial senescence," of gene pools that grow old and begin to introduce bizarre and ultimately lethal morphotypes, has been thoroughly discredited, however, and it is clear that the introduction of novelties late in the history of some groups comes not from "bad genes" but from evolutionary attempts to produce new constructions perhaps better able to survive in a changing environment.

It is tempting to continue the analogy between the life history of an individual and that of a species with regard to the variability in life spans. Members of our species, for instance, have a very characteristic longevity. If we discount the people who die early from disease and accidents, we begin to see a characteristic age of death. A small number of humans die of old age in their sixties, but more characteristically we last into our seventies. A few of us last well into our eighties, and fewer yet see a ninth decade on this earth. And then a very small number see the century mark. We celebrate these centenarians and congratulate them, and pepper them endlessly with a single question: "To what do you owe your longevity?" I have always loved the clear consensus of the responses. "I never touched a drop of alcohol," responds an old man. "A pint of whiskey a day," says another. Cigarettes. No cigarettes. Rich food. Lean food. Hard work. No work. Exercise. Plenty of rest. Lots of children. No children. Being married. Staying single. Having faith. Not cluttering your mind with that stuff. A sensible few throw in good luck. We are fascinated by these survivors, and seek to give them credit in some fashion for having had the skill to outmaneuver the grim reaper for such a long time. But even these long-lived individuals die eventually, and they usually die of the same things that catch up with most of us: heart attacks, cancer, pneumonia, infectious diseases. And so the question can be asked: Do some individuals live longer than the norm because of some special factor in their biology, or is longevity due simply to chance?

Some species have lived a very long time as well, far longer than the average life span of a species. We can determine with accuracy the longevity of many species commonly preserved in the fossil record. By applying radiometric age dating to the sedimentary rocks containing fossils of the species in question, we can (with some margin for error, of course) say with reasonable certainty that many species have lived for millions of years. When we have analyzed enough species in this way, dating their first appearance in the fossil record and their last, we can devise a table of species longevities. And like individuals, species show varying longevities. Some lived only a short time on earth, some much longer. For some groups of species in the higher taxa—the categories used to divide the species into biological units on the basis of common ancestry—we can determine characteristic durations. And just as in the case of human groups, it turns out that various groups of species have characteristic life spans that distinguish them from others. Species of mammals, for instance, usually exist for less than 5 million years before they become extinct. Species of bivalved mollusks usually last ten times as long. But even within the tightest groupings of species some last longer than others. So we're back to the question we asked about individuals: Do some species survive for a very long time because of good genes or good luck?

Darwin's Dilemma

Our view of the world and its workings is, by and large, a matter of faith. Few people would dispute the fact that the earth goes around the sun, and not vice versa; or that the moon has a back side as well as the front side so familiar to us all. But how many of us could *prove* that either of these things is true? We take so many things for granted. There is a huge economy on our planet called electronics, yet no one has ever seen an electron, and surely only a small number of people on earth can even tell you what an electron is. Most of us simply accept these facts, established by science, as true. Few people dispute the law of gravity, although no one can actually see the force of gravity, only its results (the apple falls with no strings attached). We accept the findings of modern physics. Unfortunately, our society has less confidence in the science of evolution. A recent survey indicated that only one-half of American adults believe that the theory of evolution has any basis in fact.

Scientists speak of "revolutions" in their discipline. Revolutions occur when an entirely new interpretation is offered to explain an already existing series of facts. Sometimes the revolution within a science is so great that it affects not only the science in question but many others as well. And sometimes the revolution is of such magnitude that it expands outward from the laboratories and libraries into the streets and changes the way people of all sorts view the world. When Copernicus showed that the earth does not stand still while the sun revolves around it, but in fact revolves about the sun, he started a revolution that changed our perception of ourselves and of our place in the universe. The work of Charles Darwin had the same far-reaching effects, and more than a hundred years later it still provokes us to reexamine ourselves and our place in the scheme of things.

Like all world-shaking theories, Darwin's theory of evolution is easily summarized, for Darwin's original conception rested on just two theses. First, Darwin considered that all organisms living on earth today, along with all those now extinct, descended with varying amounts of modification from one common ancestor. Second, he considered the chief agent of modification to be differences in the lengths of time various organisms survived, or, as he called it, "natural selection" acting on individual variation within populations of organisms. The biologist Douglas Futuyma considers the theory of evolution as Darwin described it in *The Origin of Species* to be the expression of two streams of thought that ran counter to views that had long prevailed.[1] First, Darwin showed that the world is and has been constantly changing, and that change is the natural order among organisms as well. Second, Darwin believed that evolutionary change has no cause (such as God's will) or purpose. He believed that material factors were sufficient to explain biological as well as other physical phenomena. With these two views, Darwin proceeded to amass evidence to demonstrate the reality of evolutionary change, to show that species change through time.

The publication of the first edition of *The Origin of Species* in 1859 was a significant scientific event. Darwin's theory (also proposed independently by Alfred Wallace, a contemporary of Darwin's) was met with both loud applause and fierce opposition. Darwin had many objects to meet, some spurious and some serious. In later editions of his great work Darwin addressed and attempted to overcome these objections. Some were easily dealt

[1]Douglas Futuyma, *Evolutionary Biology* (Sunderland, MA: Sinauer Assoc., Inc., 1986).

with, but others he was never able to answer to his own satisfaction. Perhaps the greatest problem he had to tackle was the means by which adaptive characteristics were passed on from generation to generation, for the principles of genetics were still to be discovered at the time of Darwin's death. A second problem that he could not resolve related to the nature of the fossil record. Darwin's theory required that evolutionary change take place in successive generations of creatures, through slow, step-by-step changes in form. He conceived the driving mechanism of evolutionary change to be the process he called "natural selection," or "survival of the fittest." The result of this process, the actual morphological change within an evolving lineage of organisms, should, in Darwin's opinion, have resulted in a fossil record that demonstrated slight but continuous change among successive generations. But actual fossils that demonstrated such "insensibly graded series" were rare in Darwin's time, and they remain rare to this day.

It is widely recognized that Darwin was a great zoologist, and he was clearly well versed in geology as well. He was acutely aware of the importance of evidence from the fossil record to confirm his ideas about evolution. But the fossil record, far from becoming a major source of support for Darwin, instead became a source of vexation, and he railed about it in successive editions of *The Origin of Species*. To Darwin's dismay, the fossil record—the principal record of evolutionary changes—showed very little unequivocal evidence of gradual change. To Darwin, it was the fossil record, not his theory, that was at fault. He complained that the fossil record was "poor" and incomplete, for he was sure that evidence of "insensible gradations" of change had to exist somewhere in the rocky pages of earth's history. The failure of the fossil record to support the theory of evolution was not lost on Darwin's critics.

Other critics insisted that if evolution worked as Darwin implied, there should be no "primitive" creatures, for all should have advanced, given the great length of geological time. These detractors were assuming that evolutionary change implied "progress," from primitive to advanced states. Darwin's theory implied no directed purpose to evolutionary change, and he eloquently attacked those who insisted on seeing a purpose or direction in the history of various organisms:

> But it may be objected that if all organic beings thus tend to rise in the scale, how is it that throughout the world a multitude of the lowest forms still exist; and how is it that in each great class some forms are far more highly developed than others? Why have not the more highly developed forms everywhere supplanted and exterminated the low? . . . On our theory the con-

> tinued existence of lowly organisms offers no difficulty; natural selection, or the survival of the fittest, does not necessarily include progressive development—it only takes advantage of such variations as are beneficial to each creature under its complex relations of life. And it may be asked what advantage, as far as we can see, would it be to an infusorian animiculae—to an intestinal worm—or even to an earthworm, to be highly organized. If it were no advantage, these forms would be left, by natural selection, unimproved or but little improved, and might remain for indefinite ages in their present lowly condition.[2]

Still, Darwin's central tenet was that most organisms have changed through time. But did they all change at the same rate, or did the rate of change vary? Darwin was sure that it varied, for he could point to a host of creatures that were quite similar to fossils he had seen, some from very old strata indeed. Darwin confronted this problem several times. Although he seems satisfied with the explanation he gives in *The Origin of Species*, the very fact that he repeatedly brings these "living fossils" to the attention of his readers suggests that he was not entirely comfortable with the phenomenon. He writes, for example: "In some cases . . . lowly organised forms appear to have been preserved to the present day, from inhabiting confined or peculiar stations, where they have been subjected to less severe competition, and where their scanty numbers have retarded the chance of favorable variations arising." Nevertheless, the existence of living fossils, a term that he coined, continued to puzzle him, and provided a weapon for his numerous critics to wield against him.

Rates of Evolution

Do creatures evolve at different rates? And if they do, is there a reason, or is God playing dice with the fates of organisms, as well as with the universe? Such questions were pondered by George Gaylord Simpson, one of the greatest paleontologists of all time and a founding father of the "modern synthesis," that melding of views from genetics, paleontology, and systematic biology which between 1936 and 1947 forged a new, "Neo-Darwinian" view of how evolution works. The major theme of the modern synthesis was that evolution works at the level of the population—the organisms that constitute an interbreeding group or live together in a specific habitat. Simpson considered himself to be primarily a vertebrate paleontologist, a specialist on the history of the chordates, but he was broadly trained and broadly curious, and thus

[2]Charles Darwin, *The Origin of Species and the Descent of Man*, Sixth Edition (New York: Modern Library, Random House, 1977), pp. 94–95.

he was keenly aware of the evolutionary histories of invertebrates and plants as well as of vertebrates. He had received his doctorate in 1926, and over the next ten years he published an astonishing 100 scientific papers. His *Tempo and Mode in Evolution*, published in 1944, was the most influential book ever written about paleontology. In 1953 his *Major Features of Evolution* updated and amplified many of the themes he first explored in *Tempo and Mode*. Both books deal with rates of evolution.

Simpson realized that "evolutionary rate" has a variety of meanings, and that the meaning one intends by the term should be made clear. He recognized that the rate at which gene frequencies (the proportion of various genes in DNA) changed within an evolving population would perhaps be the best measure of evolutionary rate, but in the 1940s geneticists were only beginning to master the techniques necessary to study such changes. So he turned his attention to other aspects of evolutionary rates. He described two very different phenomena that could be recognized from the fossil record: the first, which he called morphological rates, were the rates at which individual characteristics or complexes of characteristics changed within lineages of organisms; the second, which he called taxonomic rates, were the rates at which taxa with different characteristics replaced one another over time. In the simplest case, morphological rates were applied to single characteristics; Simpson used the changing dimensions of fossil teeth in a lineage evolving over some finite portion of geological time as an example of a morphological rate change.

Taxonomic rates were quite different phenomena. Comparing a species with an individual—it is born, lives, and dies—Simpson described the taxonomic rate of an evolving lineage as the rate at which species in that lineage died and were replaced by other species. In groups in which those characteristics that differentiate the species from others are changing rapidly, the taxonomic rate is high; the life span of any individual species, then, is short. Simpson's appreciation of the variability in evolutionary rates came about in part because of his original specialty: the fossil history of Tertiary Period mammals. He used examples from this source to support his work on evolutionary rates. He recognized that the duration of any given species of horse, for instance, was relatively short, and that therefore the rate of taxonomic evolution must have been high. And in comparison with the durations of species of other organisms, such as bivalved mollusks, the durations of almost all species of mammals were very short indeed.

When Simpson was first studying evolutionary rates, radiometric dating of the earth's rocks was still in its infancy; there were only the beginnings of a crude chronology for the major periods and ages of the geological time scale. Gradually over Simpson's career the number of reliable dates began to increase, and by 1953, when he published *The Major Features of Evolution*, he was able to plug in some approximate ages of the boundaries of the geological time units and then compute the longevities of various taxa. He was able to show that bivalves had evolved at a relatively slow tempo, each genus living an average of about 80 million years. Mammals showed much higher evolutionary rates —about 8 million years per genus. Simpson was intrigued by such differences, and soon had compiled an impressive list of evolutionary rates for a wide variety of taxonomic groups. Within large groups of organisms, such as families of mammals, he found what appeared to be three distinct tempos of evolution: a small group that showed an extremely fast rate; another small group that evolved at a very slow rate; and in between the majority of taxa, with an "average" rate. Among the slow evolvers were forms that seemed to be arrested in their evolution—forms that showed little or no evolutionary change over vast periods of time. These were the living fossils.

Like Darwin, Simpson and the other architects of the modern synthesis believed that evolutionary change took place gradually over long periods of time, through the cumulative effect of many tiny changes over the generations of a lineage of organisms. In this view, the longevity of a taxon, such as a species, was certainly related to its rate of morphological change, for after all, it's morphology—form plus structure—that defines a species. If the rate of morphological change is high, the changes sufficient to cause a competent taxonomist to recognize that the species has evolved into something entirely different—and hence to recognize it as a new species—will occur quickly. But a species that a taxonomist follows over time (by collecting fossils in a continuous succession of strata) is subjectively defined by the taxonomist. Organisms alive today are recognized as belonging to a species if they are capable of interbreeding: the biological species concept thus requires the tacit assumption that there is at least the potential of successful interbreeding if members of widely separated populations should meet. But what about extinct organisms? How can we be sure that if two long-dead tyrannosauruses were somehow magically brought back to life, they could successfully breed? The species concept thus undergoes a change of its own when it's applied to fossils, for its only basis is the degree of similarity between fossils. In a lineage of fossils, then, the taxonomist must decide when enough differences are present to warrant recognition of a new species.

Simpson, like Darwin, recognized that the rate of change can be affected by the two components of a species' life: the speciation event itself, when, through some isolating mechanism, a discrete population of an already existing species accumulates sufficient change (or genetic difference) to become a differentiated, "new" species; and later evolutionary change in the new species before it becomes extinct. Simpson called the first process "speciation," and later "splitting"; he called subsequent evolution within the new species "phyletic evolution." Simpson recognized that the rate of speciation and the rate of phyletic evolution were independent.

George Gaylord Simpson was very much interested in the organisms that seemed to have slow rates of evolution; he devoted a chapter of *Major Features of Evolution* to these phenomena. Yet the creatures that have the slowest rates of all, the living fossils, were only evolutionary curiosities, more embarrassments to the theory of evolution than anything else. In 1972, however, a paper published in an otherwise obscure book elevated the scientific status of the living fossils. Now they became evidence in support of an elegant new interpretation of the way new species are formed.

Punctuated Equilibrium

The seeds of scientific revolution are usually sown in the pages of scientific journals, which alight in the libraries and workplaces of scientists to take root and grow. The technical papers that spark revolution are usually preceded by learned addresses and talks; but it is the printed page that sweeps the old away.

The publications that bring about these changes are usually refereed journals: the "white" literature. It was therefore a surprise to many people that a chapter in a book, known to scientists as the "gray" literature (the published scientific output that is not first reviewed by peers), provided the forum for a sweeping reassessment of evolution. In 1972 Thomas Schopf, a paleontologist at the University of Chicago, edited a book called *Models in Paleobiology*. One of the chapters, "Punctuated Equilibria," by Niles Eldredge of the American Museum of Natural History and Stephen Jay Gould of Harvard University, presented a new way of looking at the fossil record. To say the least, this paper changed the way paleontologists thought about evolution. It has produced twenty years of clamor among evolutionists, many of whom devoutly swear that Eldredge and Gould's idea (1) is wrong and (2) had been said long before, anyway.

Eldredge and Gould went back to Darwin's objections to the poverty of the fossil record. Darwin had certainly been troubled by the seeming lack of transitional fossil forms between species, but he thought he knew the reason: "The geological record is extremely imperfect and this fact will to a large extent explain why we do not find interminable varieties, connecting together all the extinct and existing forms of life by the finest graduated steps. He who rejects these views on the nature of the geological record, will rightly reject my whole theory."[3] Eldredge and Gould had the temerity to reject Darwin's view of the geological record without rejecting his whole theory. They proposed that the lack of intermediate forms may reflect reality, not an imperfection in the geological record. Suppose, they said, that most morphological change took place very quickly, in a small, isolated population, during the speciation event itself (when an entirely new species arises, now incapable of successfully interbreeding with the species from which it arose), and that after this rapid transformational period the newly created species then underwent very little additional morphological change. No transitional forms would appear in the geological record because there had been no transitional forms.

One of the most interesting implications of this view relates to living fossils. If Eldredge and Gould are correct, a species undergoes little or no morphological change after the speciation process, and the degree of morphological change in a lineage of organisms is linked to the number of speciation events. When a great deal of morphological change is evident between the first and last appearance of a group, the group must have undergone many speciation events and evolved into a relatively large number of species. When little or no change can be seen since the appearance of the ancestor of a lineage, the group must not have evolved. The implication was clear: the living fossils, species that continue as they are for a very long time, for some reason belong to lineages that do not commonly speciate. The paleontologist Steven Stanley has this to say about living fossils: "1) they must have survived for relatively long periods of geological time at low numerical diversity [at any given time a group is represented by one or only a few species], often as the sole survivors of previously diverse taxa. 2) They must today exhibit primitive morphologic characters, having undergone little evolutionary change

[3]Niles Eldredge and Stephen Jay Gould, "Punctuated Equilibria: An Alternative to Phyletic Gradualism," in *Models in Paleobiology*, ed. Thomas J. Schopf (San Francisco: Freeman, Cooper, 1972), pp. 82–115.

since dwindling to low diversity at some time in the past."[4] Stanley points out that such a phenomenon could occur only under the punctuated equilibrium model of evolution proposed by Eldredge and Gould.

If we go along with this view, the questions we ask about the living fossils change. We no longer ask only why they have not changed for such a long time but why they have not *speciated* for such a long time.

Stanley considered that the existence of living fossils supports the punctuated equilibrium view of evolution. The fact that most creatures that we consider to be living fossils belong to groups that comprise very few species and have continued unchanged for many millions of years can be explained by the punctuated equilibrium view, but not by the gradualism espoused by Darwin and Simpson.

During the 1970s evolutionists published thousands of papers dealing with tests of the two models. In the 1980s a new interest dominated the evolution literature: the phenomenon of mass extinctions. Scientists began to understand that these mass dyings were themselves large-scale evolutionary phenomena. After a mass extinction, the biosphere of the earth was depleted of life. Because of the lack of competition among the few surviving species, the aftermath of mass extinction was adaptive radiation: the rapid formation of many new species. The nature of these new species was thus determined in large part by the nature of the species that had survived the mass extinction event.

Geological Time

Zoologists and paleontologists look at species in very different ways. To a zoologist, a species exists in the here and now; it is defined by its ability to breed with others of its kind. Time has no part in the zoologist's equation. To the paleontologist, time is the most important element: the time the species originated, the length of time it survived, the time of its extinction. Today, with our modern means to determine age, based largely on the rate of decay of radioactive remains, we can put actual numbers on these events. Though we are unable to assign what are called absolute ages to all fossil material, in many cases we can tell how many millions of years ago a species was formed, lived, and died. Geologists have always dealt in time units, even when they had no idea

[4]Steven Stanley, *Macroevolution: Pattern and Process* (San Francisco: Freeman, 1979).

how many years those units encompassed. The geological time scale in use today was developed as a relative scale, and it works perfectly well. Geologists can examine the fossils in a unit of rock and tell whether that rock is older or younger than some other rock unit.

The great units of geological time in use today—the eras, periods, and ages—were first defined not as time units but as distinctive groups of rocks that could be differentiated from other rock types on the basis of their similarity or dissimilarity. Most of the time units still used today—the Paleozoic, Mesozoic, and Cenozoic eras, the Cretaceous, Cambrian, Tertiary, and other periods— were introduced by early- and mid-nineteenth-century European geologists strictly as a means of subdividing distinctive units of rock. The Cretaceous Period, for instance, is derived from the French word for chalk, *craie*, which was used to describe the distinctive chalk beds found along the coasts of northern France, southern England, the Low Countries, and parts of Scandinavia. By coincidence, all of these very distinctive chalk rocks, such as the white cliffs of Dover and the impressive cliffs of the Normandy coast, were deposited about the same time. Their use as a unit of time, however, greatly diminished when geologists learned that a type of rock is not necessarily limited to a given time; chalk, for instance, can be (and in fact has been) deposited during any interval of time, not just during the Cretaceous Period; so can sandstone, shale, and limestone. The conditions that produce chalk are related to the environment, not to time. A scheme of geological time based on rock type was doomed to failure, as the nineteenth-century geologists quickly found out. They needed to find something that was dependent on time to establish the ages of rocks. They found it in the assemblage of plant and animal fossils commonly enclosed in sedimentary rocks. By the mid-nineteenth century it was established that large intervals of geological time could be defined on the basis of the fossils to be found within the rocks. Though the Cretaceous Period retained its chalky name, it came to be defined by the nature of its fossils rather than by the nature of its composition. Even the largest units of time are defined in this way. The three largest, the Paleozoic, Mesozoic, and Cenozoic eras, are based on the fact that the history of life shows large assemblages of creatures that allow the sedimentary rock record to be split into four large units. The oldest, the Precambrian, contains the greatest slice of time, almost 4 billion years, but is mainly devoid of fossils, for throughout this long interval life consisted mainly of single-celled organisms and scum at the bottom of shallow lakes and seas. Such soft-bodied creatures are rarely preserved as fossils, so they left only the sparsest record of their existence on earth. Only at the end of this long interval did multicellular creatures begin to emerge. The succeeding Paleozoic Era, beginning about 590 million years ago, is separated from the Precambrian Era by the appearance of numerous larger fossils, indications that the single-celled organisms had evolved into larger creatures. The Mesozoic and Cenozoic eras, too, are based on their fossil content.

Mass Extinctions and Living Fossils

The Paleozoic, Mesozoic, and Cenozoic eras were defined in 1841 by a geologist named John Phillips. These eras correspond to three great divisions of life on earth, for most of the animals and plants that are characteristic of one era are unlike those that flourished in the other eras. What's more, the fossils of those animals and plants do not gradually dwindle toward the end of an era; the end is marked by the wholesale disappearance of thousands of species, which are then replaced by largely new flora and fauna. The upper boundaries of the Paleozoic and Mesozoic eras correspond to the two largest faunal changes preserved in the stratigraphic record. At each of these major crises the majority of species then living on the land and in the sea simply disappeared.

The two greatest mass extinctions—those that ended the Paleozoic and Mesozoic eras—were of such magnitude that they were recognized by even the earliest geologists. And with increasing study of the rock record, nineteenth-century geologists and zoologists began to recognize and document lesser extinction events as well. These faunal turnovers in the rock record are so common that they persuaded many early geologists and biologists that successive catastrophes or holocausts had engulfed the world. One who was so persuaded was the French anatomist Baron Georges Cuvier. A contemporary of Cuvier's, the French stratigrapher Alcide d'Orbigny, held similar views. D'Orbigny's careful observations of the stratigraphic ranges of Jurassic and Cretaceous fossils led him to believe that entire assemblages of organisms disappeared simultaneously everywhere on earth. We now know that his interpretation of the record was faulty, but many of his data have proved to be reliable, and they form the bases for many of the Jurassic and Cretaceous time units currently in use. We know of no instance in which a single event extinguished all life, as d'Orbigny believed, and then was followed by the creation of new species. It is true, though, that the stratigraphic record is punctuated by a series of mass extinctions, of varying duration and intensity.

Extinction is the fate of every species. Since all higher taxa (genera, families, orders, and so on) are composed of groups of

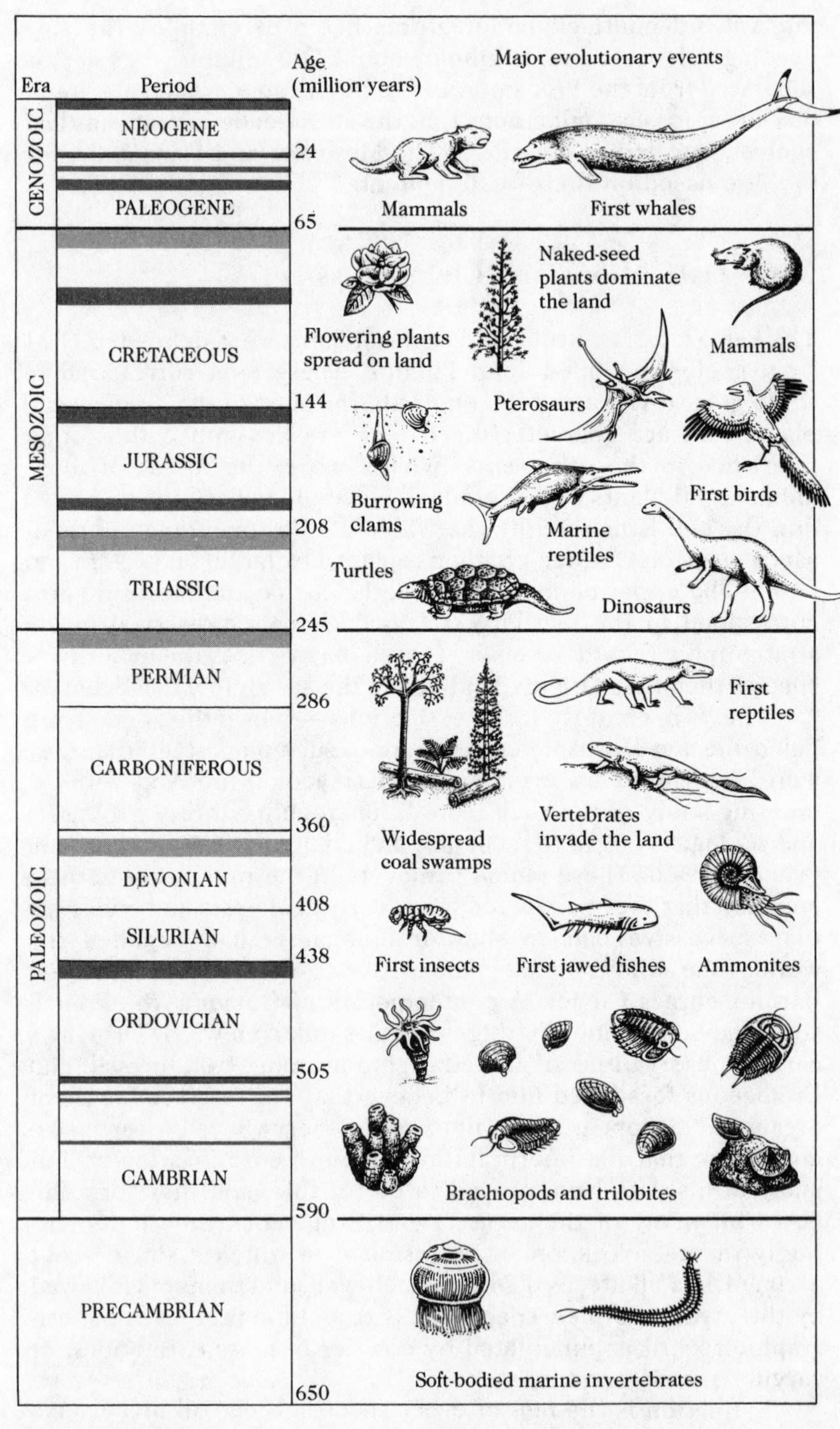

Times of the major mass extinctions shown in relation to the major evolutionary events.

species, eventually all taxonomic groups will have disappeared. Because extinction is inevitable, in any given time frame some species are going extinct. The extinction of any given species always has a cause, loss of habitat, loss of food source, the introduction of a new predator—but since such conditions can always be found somewhere, the particular extinction can be treated as a random phenomenon. This ongoing background extinction, as it is called—that is, the number of species dying out over a given interval of time—appears to have occurred at a relatively constant and sustained rate throughout all of the hundreds of millions of years since life began. Against this background, however, appear occasional periods, usually of short duration, when the rate of extinction increases far above the background level. These are the mass extinctions.

Mass extinctions have been described in various ways. According to the University of Chicago paleontologist Jack Sepkoski, they can be defined as any "interval of less than one million up to about 15 million years' duration (depending upon the magnitude of the event) during which an unusually large number of extinctions of species and higher taxa occurred."[5] Sepkoski recognizes as a mass extinction the disappearances of several major unrelated groups from a variety of habitats, terrestrial as well as marine. During such mass extinctions it's not only species that die off; even among the species that survive, the numbers of individuals may be reduced catastrophically. Usually after a mass extinction the fossil record shows a great reduction in the number of living organisms as well as of species. It is thought that at least half of all species died out in each of the major mass extinctions.

The five universally recognized mass extinctions occurred in the Late Ordovician, about 440 million years ago, when 22 percent of 450 families then living disappeared; in the Late Devonian, 360 million years ago, when a similar number of families became extinct; at the end of the Permian, about 250 million years ago (the largest mass extinction), when 50 percent of 400 families went extinct); in the Late Triassic, 213 million years ago (20 percent of 300 families extinct); and in the Late Cretaceous period, 66 million years ago (15 percent of 650 families extinct). The Permo-Triassic extinction was by far the most catastrophic: it is estimated that from 76 to 96 percent of all species on earth disappeared then. Several other mass extinctions of lesser magnitude have been described at one time or another (four or five in the Cambrian, the Toarcian Stage of the Jurassic, the Tithonian Stage of the Jurassic, the Cenomanian Stage of the Cretaceous, the Late Eocene Epoch of the Cenozoic, and the Pliocene Epoch

[5]J. John Sepkoski, Mass extinctions in the Phanerozoic oceans: A review. *Geol. Soc. America, Special Paper 190* (1982).

of the Cenozoic). Some people also argue that a mass extinction occurred during the Pleistocene Epoch—the Ice Ages—because of the disappearance of many large mammals, but as the event seems to have affected mostly land animals, its designation as a "mass" extinction is disputed.

Can mass extinctions be viewed as large-scale evolutionary mechanisms—in other words, phenomena that affect the history of life on earth? Stephen Jay Gould has followed this line of argument. In 1985 Gould proposed that the processes that shape the form and pattern of life on this planet may be acting at three disconnected levels, each with its own resulting effect on the history of life.[6] Gould called these levels tiers, or time thresholds. At the tier that Gould calls ecological time is the day-to-day life experienced by organisms as they search for food, and migrate. This is the tier at which evolution proceeds as Darwin conceived it, by the gradual accumulation of small changes over long periods of time. Such evolution in ecological time is often termed *microevolution*. Eldredge and Gould's theory of punctuated equilibrium proposes that the changes that accumulated slowly during ecological time are of so little consequence that they rarely produce speciation; speciation events occur quickly, in a burst of morphological change, followed by a long period of stasis. Whichever model is correct—and many evolutionists now think that both types of evolutionary change can occur—evolution proceeds over time spans that vary from years to centuries.

Gould's second tier of time relates to trends over millions of years, trends created by the accumulated speciation events. These trends have little to do with the day-to-day life of the species; they relate solely to the overall morphological and ecological effects of the speciation events themselves; the driving force behind these changes is sometimes termed *macroevolution*. It is at this level that we can understand the concept of background extinction. Gould believes that long-term trends in evolution can be related more to the properties of the species themselves than to the accumulated history of the large number of individual organisms that make up any given species. Such characteristics include the ability to produce new species (some species seem to produce new ones easily and often, while others produce only a few over millions of years). Some groups of species show high rates of extinction, whereas others are essentially resistant to extinction most of the time. To the latter group belong the living fossils.

According to Gould, the final tier, again discontinuous from the other two, is the level of mass extinctions. "New views on

[6]Stephen Jay Gould, The paradox of the first tier. *Paleobiology*, v. II, pp. 2–12.

mass extinctions," Gould writes, "argue that, whatever happens at the second tier, mass extinctions are sufficiently frequent, intense, and different in impact to undo and reset any pattern that might accumulate during normal times." He concludes that mass extinctions represent a phenomenon that acts on the earth's biota in a way different from the normal mode of evolutionary change. We have erred, Gould believes, in trying "to place mass extinctions into continuity with the rest of life's history by viewing them as only quantitatively different—more and quicker of the same —rather than qualitatively distinct in both rate and effect."[7]

Many scientists disagree with Gould. A few even doubt that mass extinctions have occurred at all, citing an imperfect fossil record, sampling problems, and uneven taxonomic practice. Still, the major changes in the stratigraphic record that led nineteenth-century geologists to catastrophist views must be reconciled with modern scientific theory. D'Orbigny, in describing stratigraphic boundaries he deemed to be the result of worldwide, total mass extinction, may have been close to the truth when he described these boundaries as "the expression of the divisions which nature has delineated with bold strokes across the whole earth."[8]

Some Themes

When I originally proposed the outline that has led to this book, my mission was clear: I meant to write brief natural histories of some of the well-known living fossils. As the actual writing began, however, I soon discovered that more intriguing themes than simple natural history exposition were presenting themselves. The living fossils became more interesting to me as observers and witnesses to the changes in the history of life, than as the primary subjects of biography. I found the adaptive radiations, mass extinctions, and large-scale ecological changes recorded in the stratal pages of the earth's rock record to be more fascinating by far than the lives of the living fossils themselves. Although each chapter ostensibly deals with the evolutionary history of one organism or group, other themes emerge as well: the diversification of skeletonized creatures, the conquest of the land, the rise and role of predators, and the effect of the mass extinctions. Finally, as I wrote these pages, a final theme emerged: that science is an endeavor of people. The scientists themselves provide the best stories of all.

[7]Ibid., p. 8.
[8]Alcide d'Orbigny, *Terraines Cretace* (Paris: Librarie Victor Masson, 1860).

The living fossils chosen for each of the chapters are subjective choices. There are far more living fossils than those featured here, and some of the most interesting, such as the tuatara of New Zealand or the tiny mollusk *Neopilina* of the deep sea, have been omitted. These latter two are animals which I have never seen or studied; I tried to feature those creatures with which I have some familiarity.

When I started this book I hoped my ongoing research would soon reveal an answer to a question that has long puzzled a great many people: Why do some species evade extinction far longer than most species on our planet? Charles Darwin believed that living fossils lived in habitats where they had little competition from other creatures; the American paleontologist George Gaylord Simpson considered living fossils to be ecological generalists, creatures capable of living in a wide variety of environments. Other evolutionists have speculated that the living fossils inhabit relict habitats, regions of the earth shielded from most of nature's predators and more recently evolved and more efficient competitors. But it seems to me now that there is no simple or single explanation. Like the centenarians of our society, each living fossil has its own story to tell.

2

The Advent of Skeletons

The Brachiopods

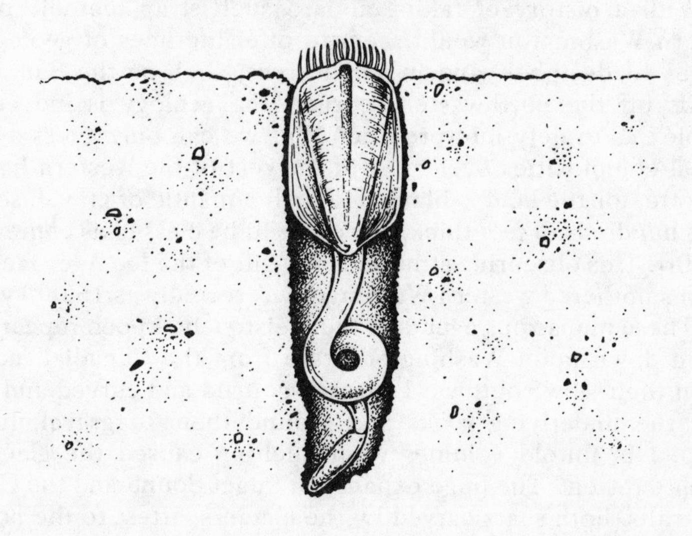

The Addy Quartzite

The State of Washington is a paradise of diversity; the dripping, sprawling forests of western Washington lap upward onto the Cascade Mountains, a chain of ragged peaks and giant, white volcanoes that bisect the state from north to south. East of the mountains the land flattens and smooths into high plains like those of the Midwest. Eastern Washington is a land of hot summers and cold winters, an expanse of prairie dotted and disciplined by rolling wheat farms.

The two halves of the state, the wet western and dry eastern sides, have in common a theme that may seem to be of no import, but that tends to irritate geologists: both sides are covered with relatively young rock deposits that tend to obscure and trivialize the normally complex geology characteristic of our continent's western shores. The western United States has had a tumultuous geological history of mountain building and microplate collision, where shards of exotic lands, originally formed far from our shores, made ancient landfall against North America's western coast, rocky flotsam driven ashore, propelled here by the engines of plate tectonics. These collisional amalgamations produced some of the most complex geological structures known in the world. Most of Vancouver Island, for instance, is composed of thick piles of rock now known to have originated in the southern hemisphere. Deposited on our coast some 100 million years ago, they created high mountain chains along the British Columbia mainland in the process of the majestic, slowly unfolding collision.

With a history of taking aboard such stray tectonic dogs, western Washington would seem to offer lifetimes of geological puzzles to decipher, and in some areas, such as the San Juan Islands off the northwestern corner, the geology is indeed so complex as to defy interpretation. But we can only guess at the crustal complexities over most of the rest of the western half of the state, for the land is blanketed with a mantle of gravel, sometimes hundreds of feet thick, so new as to be classed as contemptible dirt. This obscuring stratum is the gift of the Ice Age glaciers, which smothered western Washington as recently as 15,000 years ago. These monstrous agents of the Pleistocene Epoch repeatedly moved down upon Washington State from the Canadian north, and in their slow southward passage gouged and carved and dug away the underlying rocks and ground them to gravel, to be dropped in untold volumes when melting caused the glaciers' ultimate retreat. The huge expanse of Puget Sound and the Georgia Straits, both scars carved by the glaciers, attest to the power of these rivers of ice.

The eastern half of Washington State is similarly covered by a monotonous blanket of rock, but here the covering is of a very different nature, from a different time, even if the result is the same. Somewhere back in time, about 15 to 20 million years ago, the giant engines of plate tectonics, the mantle convective cells, changed their positions ever so slightly and began to push huge volumes of liquid magma upward. Giant forces began to tear the state asunder, and would have ripped the state in half, sending the western portion on its own tectonic journey across the Pacific to crash at last onto other shores. But, mysteriously, the process ceased several million years after it began, leaving behind untold quantities of magma from the earth's interior to mark the event.

During this aborted episode of continental splitting, giant cracks worked their way from hundreds of miles beneath the surface to the surface of the land itself. From these deep fissures liquid lava spewed forth and covered the land in waves of fire. We have all seen the horrific photos of lava flowing down on the sides of volcanoes in Hawaii and Iceland, occasionally overriding a house or church in the process. But the volume of lava that flowed across eastern Washington during the Miocene Epoch, some 15 million years ago, was enormously greater than even the gigantic flows that produced the Hawaiian Islands. Giant walls of liquid lava repeatedly swept across the land, one flow burying another, covering with basalt almost all of eastern Washington as well as large portions of Idaho and northwestern Oregon. As you drive across this land you see only endless miles of these flows, each many hundreds of feet thick, an all-encompassing blanket of magma. The best view is from a small town appropriately called Vantage, on the banks of the mighty Columbia River, which has cut a huge gorge through the layers of brick-red to brown basalts. There are a few stark testimonies to what this inferno meant to the creatures that lived in this ancient land. In the Burke Museum, on the campus of the University of Washington in Seattle, sits the cast of an ancient rhinoceros, a now-extinct species that wandered the land during the Miocene Epoch, the time of this cataclysmic tectonic upheaval in the Northwest. The cast is made of cement, which geologists had poured into a curious large hole found in the basalts of an eastern Washington wheat field. The rhino must have been running in front of the advancing wall of basalt, perhaps for many hours, for the lava poured forth at a stately pace, perhaps as fast as a person can trot, along a front many tens of miles long. And finally the exhausted rhino fell before this terrifying magma, and was covered up. Its bones and flesh quickly cooked, but the creature's body cooled the basalt immediately around it, and a perfect mold was left in the basalt. This is the hole the geologists filled with plaster. The resulting cast tells an eloquent story of a past holocaust.

Because the two sides of Washington State have been covered by relatively young deposits, one of gravel, the other of lava, there are very few places where older rocks are exposed—rocks that can tell more ancient stories. The Pleistocene gravels and Miocene basalts have largely covered, surely for as long as human life endures, the strata of past ages that make up the bedrock of the state. The ages of these rocks are measured not in thousands or even millions but in hundreds of millions of years. Only in one small corner of Washington State can ancient rocks, rocks of the Paleozoic Era and even pre-Paleozoic time, be found. It is in such rocks that the most fascinating paleontological story of all is recorded: the diversification of the metazoans, creatures like us, creatures with skeletons.

Each neophyte geologist working toward a university degree undergoes a rite of passage known as field camp, usually at the end of the senior year of training. Field camp brings together all of the theoretical aspects of geology, through the examination and mapping of actual rock bodies in their natural settings. These month-long outings usually create fond memories for the students involved, not only because they are usually held in gorgeous settings but also because of the time of life in which the students experience them. My own field camp was held in the northeast corner of Washington State, a tiny pocket spared by the floods of basalt and ravages of the glaciers. Stevens County is the one place in Washington State where old sedimentary rocks can be found.

Many of the things I saw during this field camp amazed me, and delighted me too. I had read of old sedimentary rocks and studied the fossils of Paleozoic age in my university's collections, rocks at least 200 million years old. But I had never collected a fossil even 20 million years old, let alone one ten times that age, and those I did collect were so similar to still-living creatures that they elicited little wonder, for the rocks around Seattle, the town of my birth, were young as rocks go. So it was a revelation for me finally to see truly ancient rocks, and to realize that indeed the earth was once filled with creatures very different from those alive today.

During our second week in camp our instructor loaded us into an old van and carried us northward to a small town called Addy. Addy was the site of a glass factory, for the surrounding rocks are composed of quartzite, a sandstonelike rock composed almost entirely of sand-sized grains of silica. This rock was heated to liquify the siliceous components, and the liquid silicon was turned into windowpanes.

Quickly bored by the practical aspects of glassmaking, I headed out among the rocks of the surrounding countryside, for these were by far the oldest rocks I had ever seen. The quartzites were deposited as a quartz-rich sand on the bottom of a shallow sea that existed on our continent more than half a billion years ago. The composition of the quartz-rich sediment, so clean and well sorted, could have occurred only in a setting where there were no nearby mountains; otherwise, many other kinds of mineral grains would have found their way into the sediment. The sediments show the presence of large cross-beds, sedimentary structures that are formed by the action of moving water. There must have been life of some sort on that sea bottom, but it left not a trace among the rocks I now search. Perhaps small jellyfish floated in the surface waters and tiny wormlike creatures lived among the sand grains or grazed on the algal patches growing here and there on the sandy bottom.

I try to imagine being on a sandy beach in that long-ago world, watching the sun set. The land around me looks like the pictures returned to us from the Viking spacecraft on Mars: bare rock and soil, and scattered sand dunes marching across the sterile landscape, propelled by howling winds unabated by any forests, for there are no trees, no insects, no birds, no life of any sort, except perhaps microorganisms and ragged fungi encrusting the rocks—at this time life has not yet crawled from the waters. As the light fades in the west I look to the sky's zenith, and search for some familiar constellations, but even these signposts from my time are different; I am in such an ancient land that the positions of the stars in the sky are unfamiliar. And then a rapidly brightening light in the east catches my eye. Even as the last light from the sun fades in the west, a dim brightening quickly becomes a spotlight shining upward from the eastern horizon, turning high clouds into bright crimson with a false dawn, followed by an enormous, blinding globe hurtling upward into the sky. The greatest wonder of this late–Precambrian Era world, more than a half-billion years ago, would have been the first glimpse of the moon, at that time perhaps twice as close to the earth as it is today. It appears as a giant planet rising, blinding white with its reflected sunlight, and moving across the sky much more quickly than it does today. I feel the land beneath my feet tremble slightly with the sway of a small earthquake, for minor tremors shake the earth as the passing moon stretches grasping gravitational fingers into the Earth's crust. The night has now turned into a dimmer but serviceable day in the bright moonlight, and I cast a stark shadow on the sand as the stars above me are snuffed out. And then I hear a murmur of water, and all thoughts of the moon are put aside. I look to the direction of this low rumble, out to sea, and in the moonlight see an approaching wall of water. It is a giant standing wave, moving toward me rapidly. In a flash I remember seeing the tidal advance in the Bay of Fundy, in my world, and I remember marveling at the rapidity and power of that tide's advance. But the tidal rise I now see dwarfs the tidal changes of my world; I am now facing a sixty-foot tidal change, and I understand why the beach I stand on is many miles wide. The closeness of the moon in this world pulls at the earth's oceans with its gravitational force, creating monstrous tides that race across the shallow seas, tearing at the sea bottom, shifting untold tons of sediment, and ultimately making the rise of life very difficult. Even the atmosphere discourages life at this time, for its oxygen content is far lower than that of our world.

I kicked an empty pop can with my thick field boots as I walked along the country road near Addy, surrounded by the roadside sandstones, vestiges of that long-ago world. I was walking stratigraphically upward in these sediments; they lie at an angle, tilted about 30 degrees from their original horizontal. As I walked northward along the road I was thus going up through time, into ever higher and thus younger beds of these sandstones.

With each step I passed upward through thousands of years of time; with my quarter-mile hike along the road I had traversed several millions of years among these buff-colored sandstones. I was somewhat disappointed. I was training to become a paleontologist and disdained geological phenomena not associated with fossils (I have become more tolerant since then); the quartzites shone brilliantly in the sun but were empty of life. I continued to walk along the road amid the beer cans and scrubby grass, windblown by the vast semis and farmers' trucks screaming by, stooping once in a while to examine another piece of rock littering the side of the busy highway. I finally entered a turnout, hunger now gnawing, to find a rocky cliff of the same sandy quartzite. A mound of rocky slabs formed a hill of talus at the back of the turnout, amid piles of roadside litter just beneath the "no dumping" signs. I was alone amid these rocks, my fellow students left far behind amid the joys of industrial glassmaking. I climbed the pile of talus and unshouldered my pack. A can of root beer and a flattened peanut butter and jelly sandwich served for lunch as I idly looked at the loose slabs, waiting for the shot of sugar to shake the fatigue from my legs. I had been lulled by the walk and the endless slabs of sandstone showing nothing but featureless bedding planes. The small slab now in my hands thus elicited no immediate response. I stared with unseeing eyes at the small oblong shell and tossed the rock before the message from my eyes finally burned through into my brain. The small rock followed a beautiful ballistic arc down the talus slope as I realized that I had just seen an unmistakable announcement of life. I tracked the path of the slab through the air, trying to determine its landing place among hundreds of similarly colored sandstone slabs. I picked up another piece and saw more shells, amid even more wondrous fossils. I saw the heads of large trilobites, looking something like large crabs yet very different, fossils with segments and strange crescent-shaped eyes unlike anything now living. I was surrounded by fossils, sitting atop a teaming graveyard, a joyous assemblage announcing that after 3 billion years skeletonized life had arrived. I was sitting on the base of the Cambrian System, the start of the Paleozoic Era, the beginning of the Phanerozoic, the time of life.

The Base of the Cambrian

The Addy quartzite preserves one of the great miracles of the geological record. It is a transition that can be found at thousands of places around the world. Over very narrow stratigraphic thicknesses, in places no more than a foot or two, completely barren strata, which at best contain algal mats or tiny tube-shaped fossils, are replaced by rocks with numerous fossils. And, as at Addy, the fossils usually belong to only two kinds of life. The most conspicuous are the fossilized trilobites, creatures now extinct and most closely resembling such arthropods as the pill bug and sow bug. Trilobites make up by far the greatest part of the lowest Cambrian fauna. Interspersed among the abundant trilobite fossils are a few small shells that look for all the world as if they came from some species of clam. But closer examination shows that these small, rare valves have very distinct differences from clams. First, the shells seem to have no hinge structures to hold them together, such as the teeth and socket structures found on most clam shells. And second, the interiors of these small shells, if they are well enough preserved, show the impressions of numerous small muscle scars. Clam shells close through the coordinated action of two muscles, sometimes only one. The animals that lived in those small shells had numerous muscles. These tiny shells are the first record of the brachiopods.

The history of life thus seems broken into two great divisions: a long period, now known to approach 3 billion years in length, characterized by no creatures that left skeletal fossils, followed by a shorter, 600-million-year period typified by numerous fossils. The transition between these two great divisions seems, in most stratigraphic sections, to be astonishingly abrupt: as at Addy, strata barren of fossils are overlain by strata rich in shells and the exoskeletons of trilobites.

The seemingly sudden appearance of skeletonized life has been one of the most perplexing puzzles of the fossil record. How is it that animals as complex as trilobites and brachiopods could spring forth so suddenly, completely formed, without a trace of their ancestors in the underlying strata? If ever there was evidence suggesting Divine Creation, surely the Precambrian and Cambrian transition, known from numerous localities across the face of the earth, is it.

The apparently sudden appearance of trilobites and brachiopods was surely trying for Charles Darwin. He devoted a section of one of his chapters in *The Origin of Species* to this enigma. Darwin's problem was quite clear. According to his theory of evolution, creatures as complex as these arthropods and brachiopods required many millennia of evolution to reach such a stage of development. Yet, in section after stratigraphic section, no intermediate forms were ever found. Instead of strata containing fossils as complex as trilobites lying immediately atop strata barren of fossils, Darwin expected to find intervening strata showing fossils of increasing complexity until finally trilobites appeared:

he expected to see the slow, step-by-step progression to the trilobite shape preserved in the fossil record. Darwin described this paradox in the following fashion:

> There is another and allied difficulty, which is more serious. I allude to the manner in which species belonging to several of the main divisions of the animal kingdom suddenly appear in the lowest known fossiliferous rocks. Most of the arguments which have convinced me that all the existing species of the same group are descended for a single progenitor, apply with equal force to the earliest known species. For instance, it cannot be doubted that all the Cambrian and Silurian trilobites are descended from one crustacean, which must have lived long before the Cambrian age, and which probably differed greatly from any known animal. Consequently, if the theory be true, it is indisputable that before the lowest Cambrian stratum was deposited long periods elapsed, as long as, or probably far longer than, the whole interval of the Cambrian age to the present day; and that during these vast periods the world swarmed with living creatures.[1]

The scientists of Darwin's day had no real idea how old the earth was. Some of his contemporaries estimated that the base of the Cambrian was no more than 60 million years old, and that the earth consolidated only 200 million years ago; Darwin was thus faced with but a short span of time to accomplish a great deal of evolution. But even these age constraints were not the fundamental problem. Over and over in *The Origin of Species* Darwin returns to the fact that—at least in his time—no fossils were known from the thick succession of rocks underlying the Cambrian fossil-rich rocks: "To the question why we do not find rich fossiliferous deposits belonging to these assumed earliest periods prior to the Cambrian system, I can give no satisfactory answer."

Until he died, Darwin was convinced that an enormously long history of life preceded the appearance of the basal Cambrian fauna. Many of his contemporaries held very different views. Sir Roger Murchison, one of the great pioneering geologists of the nineteenth century and a man who devoted most of his long professional career to studying rocks of earliest Paleozoic age, was convinced that the strata underlying the lowest Paleozoic rocks (which he called the Silurian rather than the Cambrian System) held no fossils for the simple reason that life had not yet appeared. In Murchison's and many other geologists' minds, the sudden appearance of life at the base of the Paleozoic Era her-

[1]Charles Darwin, *The Origin of Species and the Descent of Man*, Sixth Edition (New York: Modern Library, Random House, 1977), p. 252.

alded the origin of life. Murchison had a simple term for the rocks beneath the Cambrian strata: he called them "Azoic," or devoid of life. Murchison's great rival was Adam Sedgwick, the geologist who first defined and named the Cambrian System. Sedgwick was equally puzzled by the "sudden appearance" of fossils in the strata he studied. He named the assemblage of trilobites and brachiopods from lowest Cambrian strata "primordial fauna." Sedgwick had a new idea about the apparent absence of life beneath Cambrian strata: he concluded that these underlying sediments had undergone metamorphism—a pronounced change in their composition. Fossils had originally been there, he decided, but they were destroyed when the rocks became heated. This argument did not hold up long: Sedgwick and others soon discovered successions of strata, both with and without primordial fauna, in whose sediments no evidence of metamorphism could be detected.

A more reasonable explanation for the apparently sudden appearance of the trilobites and brachiopods was proposed by the great American geologist C. D. Walcott, who suggested that the sediments representing the time interval immediately before the basal Cambrian Period either had been removed by erosion or had never been deposited. He called this missing interval the Lipalian Period. He thus considered than an intermediary fauna had once been present but was not preserved in sediment:

> The apparently abrupt appearance of the Lower Cambrian fauna is therefore explained by the absence on our present land areas of the sediments, and hence the faunas, the Lipalian Period. This resulted from the continental area being above sea level during the development of the unknown ancestry of the Cambrian Fauna.[2]

Walcott had reasonable evidence to back up his claim. One of the most striking observations about the latest Precambrian rocks and the earliest fossil-bearing strata above them was that they were composed of either clean, quartz-rich sandstones or quartzites, which form when quartz-rich sands are slightly heated or buried. In either case the presence of such rocks gives a valuable clue to the environment at the time they were deposited. Quartz-rich sandstones are often found on beaches or on the bottoms of shallow seas which have been subjected to extensive reworking by waves and currents. To the geologist, deposits of this sort also signal a long-term rise in sea level, for the slow encroachment of a

[2]C. D. Walcott, *Smithsonian Inst. Misc. Coll. V.57* (Washington, D.C., 1910), pp. 1–16.

rising ocean creates conditions that produce the reworking, winnowing, and sorting of sand grains which ultimately create quartz-rich sediments. The occurrence throughout the world of quartz-rich sands or quartzites at localities of latest Precambrian or earliest Cambrian age is strong evidence that the sea level was rising everywhere during this critical junction in the history of life. Walcott knew that times of lowered sea level produce little sediment that is actually preserved and incorporated in the stratigraphic record. It thus seemed reasonable to him that a period when no sediments were deposited or preserved, the "Lipalian interval," hid the ancestors of the primordial, Cambrian fauna. But as more stratigraphic sections of this critical period were studied, it became clear that Walcott's Lipalian interval had not existed, and well into this century the mystery remained unresolved.

The apparently sudden appearance of skeletonized life at the base of the Cambrian system cried out for explanation. Was the observed phenomenon real? Did life really originate at the base of the Cambrian, some 560 to 570 million years ago, as Murchison proposed? The first question was easily disposed of, for as early as the mid–nineteenth century geologists working in Canada and other parts of the globe found and reported large globular bodies that looked much like layered mats, which they interpreted as the remains of interlayered algal cells and sediment. These large calcium-rich structures, called stromatolites, appear to have formed when slicks of blue-green algae were covered by a thin layer of sediment, which the algae then colonized by sending filaments upward. The resulting sediment-algal constructions grew to be many feet in diameter, and some towered several feet above the sea bottom. These structures were found in strata far older than the trilobite-bearing beds, and are now known to have been the most common organisms on the earth during most of its history. For most Precambrian time they flourished in shallow seas around the globe. Then, mysteriously, they began to disappear, not long before the first appearance of the early Cambrian trilobite and brachiopod fauna. We now have reasonably precise radiometric dating of the disappearance of the stromatolites. Somewhere about 800 million years ago their numbers began to diminish, until very few were left to fossilize. Several explanations have been offered as to why the stromatolites virtually disappeared. Perhaps the climate changed throughout the world, for there is evidence that an extensive glaciation occurred in late Precambrian time. A more reasonable explanation suggests that the stromatolites succumbed not to weather but to grazing: they became the favored foodstuff of a new type of organism, and were literally eaten off the bottom of the sea.

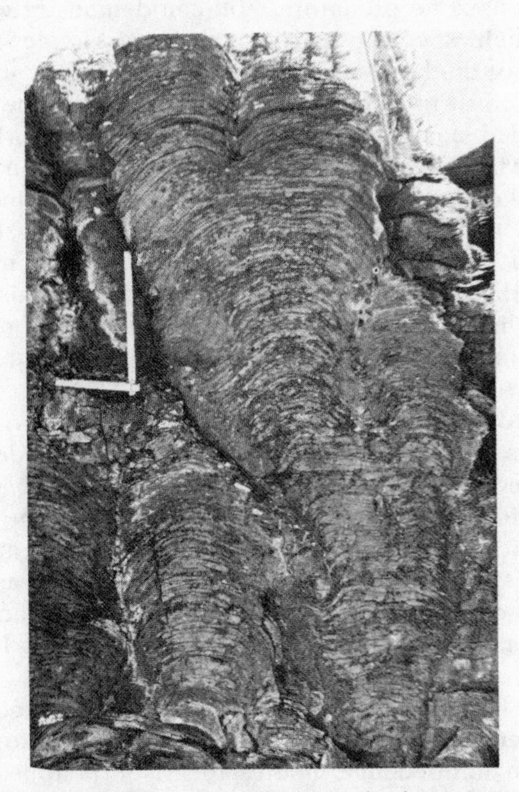

Living stromatolites at Shark Bay, Australia (top), *and similar organisms from the Precambrian Period* (bottom). (*Courtesy of P. H. Hoffman.*)

Stromatolites exist today, but only in a few very special environments. The most famous locale is a dry, hot place on the west coast of Australia known as Shark Bay. Here, in the hot shallow water, large hemispherical heads of interlayered algae and sediment exist in large numbers, sometimes reaching four to five feet in diameter. If they are cut apart, they show shapes and structures identical to fossil forms that are as much as 3 billion years old. Of all living fossils on earth, these strange structures still living in Shark Bay are by far the oldest. They exist in this place for a simple reason. Try as you may, you will find virtually no other creatures living among the stromatolites of Shark Bay. This is quite an anomaly: in similar settings around the tropics, such shallow bays are usually home to huge populations of plants and animals. But Shark Bay, through geological accident, has a very restricted pattern of water circulation. That fact, coupled with the searing heat and low precipitation of this region, has caused the water of the bay to become and remain supersaline. It is a virtual brine bath, poisonous to almost all marine animals. The stromatolites of Shark Bay exist simply because they are the only species there. They have no predators. You can demonstrate the significance of their special environment by a simple experiment. Uproot one of the large stromatolites of Shark Bay and transport it down the coast to a spot of normal salinity. Put the large block of interbedded algae and lime into the water at its original depth and watch what happens. Over the following days and weeks the normal algal grazers of the sea, the limpets and echinoderms and crustaceans of the modern world, will have a feast, removing every bit of living tissue from the top of the stromatolite. Although the three-dimensional structure of the organisms will still exist, all living matter will have been removed from the top, growing surface; for all intents and purposes, the stromatolite is now dead.

Stromatolites grow slowly, far too slowly to survive the grazing of modern creatures adapted to eat algae. Their design is now obsolete. And there is reasonable evidence that it became obsolete long before now—long, even, before the advent of the trilobites. Most scientists believe that the virtual disappearance of the stromatolites, about 700 to 800 million years ago, was due to the evolution of organisms of a new type—the metazoans, or multicellular organisms, of which our species is but a late-arriving member.

There is little argument that life has been present on the earth for more than 3 billion years. Why, then, did it take so long for animals with multicellular bodies to arise? It appears that for most of earth's history life has consisted of very simple creatures, most of them single-celled forms without a nucleus, simple life forms called procaryotes. The emergence of cells containing a nucleus and other organelles characteristic of higher animals and plants was thus a relatively late event in the history of life, occurring at most about 1.4 billion years ago. The best guess is that the evolution of metazoans, multicellular creatures, occurred about 800 million years ago. Their emergence coincided with the dwindling and virtual disappearance of the stromatolites, which were literally eaten out of existence by these newly evolved but diminutive grazers.

The earliest metazoans were probably small, perhaps a few millimeters long, and may have looked like tiny flatworms. There is virtually no chance that such creatures could leave a fossil record, for they would have had no skeletons. For the next 200 million years they must have evolved and diversified, but without growing larger or developing skeletons. We can imagine an army of wormlike creatures, some with segmentation, some without, living in and on the sediments of the world's oceans during the interval from 800 to 600 million years ago, at first grazing peacefully on algae, then perhaps developing a more carnivorous means of getting food, for some of these groups assuredly must have turned to others of their kind for food.

We have few windows into this world. This entire panopoly of evolution, when the major phyla of animals and plants were developing the body plans characteristic of their group—this most interesting moment in the evolution of life occurred among tiny creatures that left few clues to their identity. Only here and there can we get a glimpse of this most crucial of times in the history of life.

Until almost 1950 the absence of metazoan fossils older than Cambrian age continued to puzzle evolutionists and earth historians alike. Other than the remains of single-celled creatures and the matlike stromatolites, it did indeed look as if larger creatures had arisen with a swiftness that made a mockery of Darwin's theory of evolution. This notion was finally put to rest, however, by the discovery of the Ediacarian and Vendian fossil faunas of latest Precambrian age.

The Ediacarian fauna, as it came to be known, was discovered in 1946 in a dry, nearly lifeless part of the Australian outback north of Adelaide. There, while prospecting for metals, an Australian government geologist named R. C. Sprigg discovered the impressions of creatures that looked like jellyfish preserved in sandstones of late Precambrian age. These extraordinary discoveries were soon followed by even more spectacular finds: the same sandstones yielded the remains of creatures resembling worms and a variety of soft corals. All of these fossils were assigned to still-living groups, such an anneled worms and coelenterates. But later study cast doubt on the affinity between these ancient remains preserved in sandstones and living creatures of today; the great German paleontologist A. Seilacher, of Tübingen University, has even gone so far as to suggest that the Ediacarian fauna has no relationship whatsoever with any currently living creatures. In this view, the Ediacarian fauna was completely annihilated before the start of the Cambrian fauna. Other scientists take a less extreme view. The most important characteristics of the Ediacarian fauna are their substantial size—some were more than a foot long—and their lack of any skeletal hard parts. That they have been preserved at all is something of a mystery, for the fossil record shows few examples from creatures with no skeletal elements. Since Sprigg's discovery the Ediacarian fauna has been identified on other continents. Its age, though variable, ranges between about 650 and 600 million years.

Soon after the discovery of the late Precambrian Ediacarian fauna, paleontologists in the Soviet Union and North America discovered another assemblage of Precambrian creatures, slightly younger than the Ediacarian fauna. Intensive searching of strata immediately underlying the well-known basal Cambrian deposits in the years between 1950 and 1980 showed that the larger skeletonized fossils (such as the trilobites and brachiopods) that supposedly appeared so suddenly were in fact preceded by skeletonized forms so small as to be easily overlooked by the pioneering geologists. Most of these forms, now known as the Tommotian fauna, are composed of small tube and conical shells. They are evidence of an extensive fauna consisting of worms and mollusks at least 600 million years old. Virtually none of these forms showed any apparent phylogenetic connection with the Ediacarian fauna. But they do seem to be the immediate ancestors of many of the creatures found as fossils in lowest Cambrian deposits.

The discovery of the "Tommotian" fauna can be credited as much to new techniques as to new resolve. Early paleontologists searched assiduously for fossils in Precambrian strata, but they searched with their eyes alone. They would slowly walk the rocky outcrops or on hands and knees scan the surface of the rock in search of fossils; and they found nothing. The Tommotian fauna was discovered largely through the breakdown and preparation of these same rocks in the laboratory. Using a laborious sequence of physical disaggregation and chemical dissolution in various acid baths, followed by screening and microscopic examination, geologists discovered the variety of tiny tubes and small skeletal elements that had to have been secreted by metazoan animals about 600 million years ago. The long-accepted theory of the sudden appearance of skeletal metazoans at the base of the Cambrian was incorrect: the basal Cambrian boundary marked only the first appearance of relatively large skeleton-bearing forms, such as the brachiopods and trilobites, rather than the first appearance of skeletonized metazoans. Darwin would have been satisfied. The fossil record bore out his conviction that the trilobites and brachiopods appeared only after a long period of evolution of ancestral forms.

Factors Leading to the Cambrian Diversification

Many factors apparently went into the adaptive radiations of metazoans near the Precambrian-Cambrian boundary. At that time, about 570 million years ago, the levels of oxygen in the atmosphere and dissolved in sea water, which had been rising slowly for well over a billion years, finally amounted to about 8 to 10 percent of today's values. The size of the Precambrian animals had long been limited by the amount of oxygen available to them. At low oxygen levels, single-celled animals were the most efficient forms, because the large size of their surface areas in relation to their volume allowed oxygen dissolved in the sea to diffuse easily into all parts of their bodies. As oxygen levels increased, larger bodies became practical. The rise of oxygen also greatly enhanced the biochemical pathways that led to precipitation of skeletal hard parts. Another factor that promoted the growth and diversification of metazoans was a rise in sea level throughout the world. Late in Precambrian time, for the first time in many millions of years, large continental areas became flooded. Thus vast new shallow seas became available for colonization by marine creatures. Global temperatures may have been rising as well, for the period between 700 and 800 million years ago was a time for the most extensive glaciations the earth had known. When these glaciers finally receded, the sea level rose again, and the temperatures of both air and seawater rose as well. Like an increase in the oxygen level, a rise in the temperature of the sea greatly increased the ability of animals to form skeletons.

The latest part of the Precambrian Era was a time of increased biotic diversity as new species were formed. But it may not have been only species that proliferated; the fossil record of this time suggests that the number of animal and plant individuals increased greatly as well. The biomass on the earth—the actual volume of living creatures—increased radically at this time, and this increase may have had as much to do with the rise of higher animal and plant life as any other factor. For the first time in the history of the earth, life became so commonplace that the food resources available to many ecosystems changed dramatically.

Much of the investigation of latest Precambrian faunas focuses on the evolution of larger creatures, mostly living on or in the sea bottom. But a much more significant event may have been taking place among the plankton. Just as grass species make up the base of the food chain for many terrestrial ecosystems, plankton—the microscopic assemblage of celled animals and plants floating in the surface regions of the ocean—provide the base of the food chain for marine systems. If this basic resource increased in diversity and biomass in the late Precambrian, as many scientists think it did, it would have had an enormous impact on the forms that lived at the bottom of the sea, for the increasing resources provided new opportunities and incentives for evolutionary innovation.

Let us imagine a late-Precambrian environment, the bottom of a shallow sea, perhaps 20 feet deep. For 50 million years this water has been filled with plankton. The warm sun overhead causes the single-celled plants floating in the top several feet of the shallow, warm sea to grow quickly, releasing increasing volumes of oxygen into the atmosphere. Untold numbers of these microscopic plants grow explosively until they have exhausted the sea's inorganic nitrate and phosphate nutrients. Living among the floating plant cells are large numbers of microscopic protozoans and metazoans, feeding greedily on the phytoplankton. But as the nutrients necessary for plant growth disappear, the phytoplankton ceases to grow, and the remaining stocks are soon consumed by the voracious zooplankton. When most of the phytoplankton has disappeared, the zooplankton starves, dies, and falls to the bottom below.

This is the scenario that is played out every year in our world, usually twice a year in temperate waters. The cycle may have commenced in the late Precambrian. Such a system provides a rich and abundant source of food for two types of larger bottom-dwelling organisms: filter feeders, organisms that catch living plankton by straining the individual cells from the water around them; and deposit feeders, organisms that ingest the sediment of the sea floor and with it the newly dead plankton that has fallen there. Most of the larger metazoan animals of the earliest Cambrian were of one sort or the other.

Late in the Precambrian the increasing food resources provided by the plankton may have become a spur for evolutionary change. The tiny wormlike creatures of that time began to exploit this resource, at first, perhaps, simply by ingesting sediment, hoping to catch organic material among the inorganic grains of sand and mud. Digestive systems changed to facilitate intake of ever-larger volumes of sediment, for the successful use of this resource became a numbers game: the larger the volume of sediment processed in a day, the more organic material available to the creature. Very quickly a wide variety of creatures evolved to exploit this resource. By far the richest region of organic material was the sediment's surface. Many organisms developed ways to ingest the organic material off the sediment's surface, by first roiling the sediment surface and then scooping the agitated material along the very top.

The evolution of anatomical structures capable of using the plankton that fell from the upper waters was probably a fairly simple matter, for deposit-feeding organisms have only to ingest sediment and then extract the organic matter from it as it passes through the gut. Exploiting the living plankton was much more difficult, for the concentration of living material in water was usually far lower than the concentration of dead material on top of the sediment. To exploit plankton as a food source directly from the sea required the evolution of some filtering device, and a wide variety of such methods evolved. Some creatures, such as sponges, evolved specialized cells, called choanocytes, equipped with long, beating flagellae. The beating of these cells creates a water current that is drawn into the sponge. Microscopic plankton is carried into the sponge in these incoming currents and is strained from the water along specialized sites surrounding the beating flagellum. This system is not especially efficient, however; an average sponge must draw in many hundreds of times its own body volume of sea water each day to get enough nutrients to sustain life.

A better solution was evolved by a small group of wormlike creatures. Partially buried upright in the sediment, these creatures evolved a lacy, fanlike structure covered with beating cilia around their mouths. Coordinated beating of these cilia created a water current, which passed through the stretched-out tentacles; the cilia snagged food material from the water current and passed it to the mouth. Although this system allowed direct exploitation of the plankton, it was fairly inefficient because water tended to pass around the tentacles rather than through them. So a third solution was produced. The same upright wormlike creatures, today called phoronid worms, evolved an ingenious way to increase filtering efficiency: they enclosed the lacy, tentacular feeding apparatus, called a lophophore, within a proteinaceous shell. The tentacles were then stretched across the middle of the shell, so that all water drawn into the shell had to pass through them. The system was amazingly efficient, and it had the additional benefit of protecting the vulnerable feeding tentacles of the lophophore from attack by newly evolving carnivores. With the

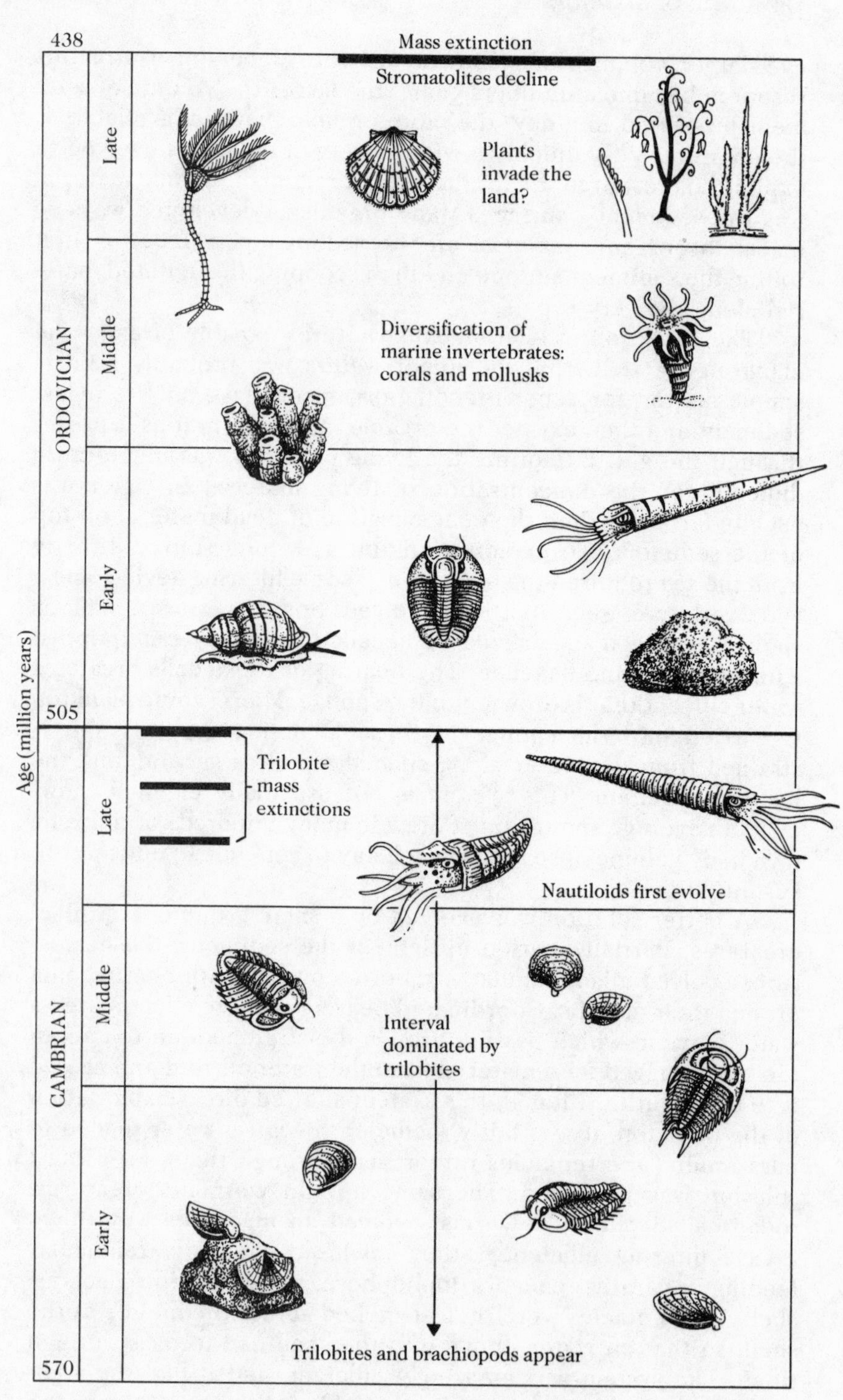

Major events of the early Paleozoic Era. The trilobite and brachiopod fauna appear near the base of the Cambrian Period.

evolution of this shell system, first developed to facilitate feeding rather than for protection, the brachiopods came into being.

The development of a thin proteinaceous shell to enclose the feeding lophophore structure must have been a wildly successful innovation, because many new species that used this system soon evolved. Just as each species that becomes extinct is represented at the end by some final individual, last of the stock, each new species begins with one individual, the very first creature to introduce an innovation. There must have been some moment when the first phoronidlike creature, through a chance mutation, created the first shell that completely covered the lophophore. Perhaps for millions of generations the shell portion had been virtually complete, lacking only some small extension that would close it. Perhaps this new method of filtering plankton gave that first complete creature markedly more energy than the others of its species and permitted it to grow faster; and if this innovation was fixed in its genes, its young, too, would have been superior to members of the species that lacked the complete shell. Very quickly, probably, natural selection would have favored those individuals in the population with the complete shell system, so that in not too many generations the new shelled forms would have proliferated at the expense of their less efficient brethren. A new species was being born, a species that would revolutionize the marine ecosystems throughout the Paleozoic Era. It is these first experiments in shelled, lophophore-bearing creatures that are found in the Addy quartzite, and at many other basal Cambrian localities around the world.

The Ohio Valley

I am once again stretched out on a rocky outcrop, enjoying lunch on a fine spring day in 1977. The lush green of the Ohio Valley seems like a graceful apology from nature, for the winter of 1977 was one of the coldest on record; the temperature in normally mild central Ohio did not rise above freezing for more than two months. Amid the bursting shoots and leaves of this late-April afternoon, the winter seems like a bad dream.

I watch manic tadpoles scurrying through the sluggish stream before me. They dash about like bumper cars, their long black tails cleaning off a fine silt cover from the underlying rock like so many tiny brooms, occasionally unveiling organized shapes in the underlying limestone, shapes that speak of past life, of sea bottoms rich in color and vitality, now an eroding memory. This tiny tributary I watch, home to tadpoles and grave site of the Paleozoic Era, is but a capillary among the innumerable streams that feed the Ohio River, and through the unfolding leaves I can dimly see that mighty river in the distance as it bends southward toward Cincinnati. I have come to this place with the students in a class I teach, one of my first groups of students ever, for I am in my first year of teaching at the Ohio State University. I have brought the students here to measure strata, practice constructing geological maps, and collect fossils from the flat-lying limestone strata that the Ohio River has excavated over the many long years since the glaciers last retreated from this region. I look in wonder at the stratum on which I sit. I have come from a land where even the young Ice Age deposits are tilted, where the ages of the rocks can be counted in only thousands of years, and now I sit on a stratum that is still nearly as horizontal as it was when it was deposited as sediment on a seabed more than 400 million years ago. The Midwest has clearly escaped the fury of mountain building and the collision of exotic terrains; during its geological history the land has been, at most, gently warped. It seems to me a place old beyond measure.

Flat-lying strata of immense age are surprise enough to a West Coast boy. But even more astonishing is the content of these rocks. Idly I pick up a loose piece of the buff-colored limestone. It is covered with fossils. Everywhere I look I see the remains of life: shells, calcareous branches, coiled spiral designs. The thick strata that make up the American heartland and that yielded the rock I hold are among the richest known concentrations of fossils of any age in the world.

The tiny creek we have chosen as our lunch stop has cut through several tens of feet of this Ordovician-aged limestone. Because of the extraordinary profusion of fossils to be found here, an international governing body of geologists has designated the area to serve as a stratotype, or worldwide reference section, of Lower Paleozoic times. This rich area has been a breeding ground of American paleontologists, from early giants such as E. O. Ulrich to contemporaries of mine. Many people have been drawn to these rocks as children, perhaps armed at first with claw hammers and makeshift chisels, to collect fossils. And many of those children grew up to make a career of studying them.

As the afternoon progresses we find a particularly well-exposed bank and begin collecting. The soft countryside is soon filled with the sounds of our industry, the crack of hammer against rock or chisel, and the cries of joy or moans of disappointment as pried fossils either pop out of their entombing sediment whole or emerge cracked and useless, a keeper gone sour and chucked. I hide a smile as I watch this industry, for these college students are not far removed from more childish pursuits, and they appreciate this type of work for the childish fun and wonder of it. I am reminded of the seven dwarfs, singing with industry as they make their way to the mountain.

It's late afternoon when we finish, our packs bulging. We have laid out a representative array of the creatures from this place, or at least the creatures that lived here so many millions of years ago, when the Ohio River was still but a dream and the rolling countryside was covered by a warm, shallow sea whose bottom was rich with life. The fossils we have collected tell a story of that place, giving undeniable clues to the nature of that sea. Its bottom, covered with fine, lime-rich mud, must have been less than 100 feet deep. There were no fish, but the number of crustacean-like creatures was vast. Most important of these creatures were the trilobites, descendants of those same jointed creatures found in such abundance at the Addy quartzite of eastern Washington, but forms vastly different from those pioneering species of the earliest Cambrian. The early Cambrian forms were profusely segmented, with almost wormlike bodies and curious crescent-shaped eyes; the fossil trilobites we find in these Ohioan strata, deposited more than 50 million years later than the Addy quartzites, are squatter and they have far fewer body segments. Some are found rolled, like pill bugs of our modern world. Many show spines, festooned like spears extending outward from a suit of armor, testament to the need for defense against predators. Evolution worked its magic among the trilobites during the time between the early Cambrian deposits and the strata we are studying now. By the end of the Cambrian Period, about 500 million years ago, the time of trilobites had largely come to an end. Even though this group would linger on until the end of the Paleozoic Era, they would never again have the diversity they attained during the Cambrian Period.

I examine other fossils proffered by eager hands. I see an assortment of filter-feeding creatures, such as twiglike bryozoans, and a variety of small corals. Here and there rarer treasures are disclosed, such as the occasional crinoid, a relative of the sea stars and urchins of our world, and the nautiloid cephalopod, the dominant predator of this long-lost Ordovician world. But most of the fossils I am shown look like small clams. There are hundreds of them; they virtually pave the limestone surfaces, they are a richness of this lost world: if they were gold, we would all be Croesuses. They are familiar to me from my work at Addy, and from another place as well. I know these creatures not only dead but alive. These brachiopods, the same sort of creatures I collected at Addy, are among the first wave of larger animals to have evolved at the start of the Cambrian. But unlike the earliest Cambrian rocks I have seen in eastern Washington, in which brachiopods are rare, these later rocks of the Ohio Valley yield

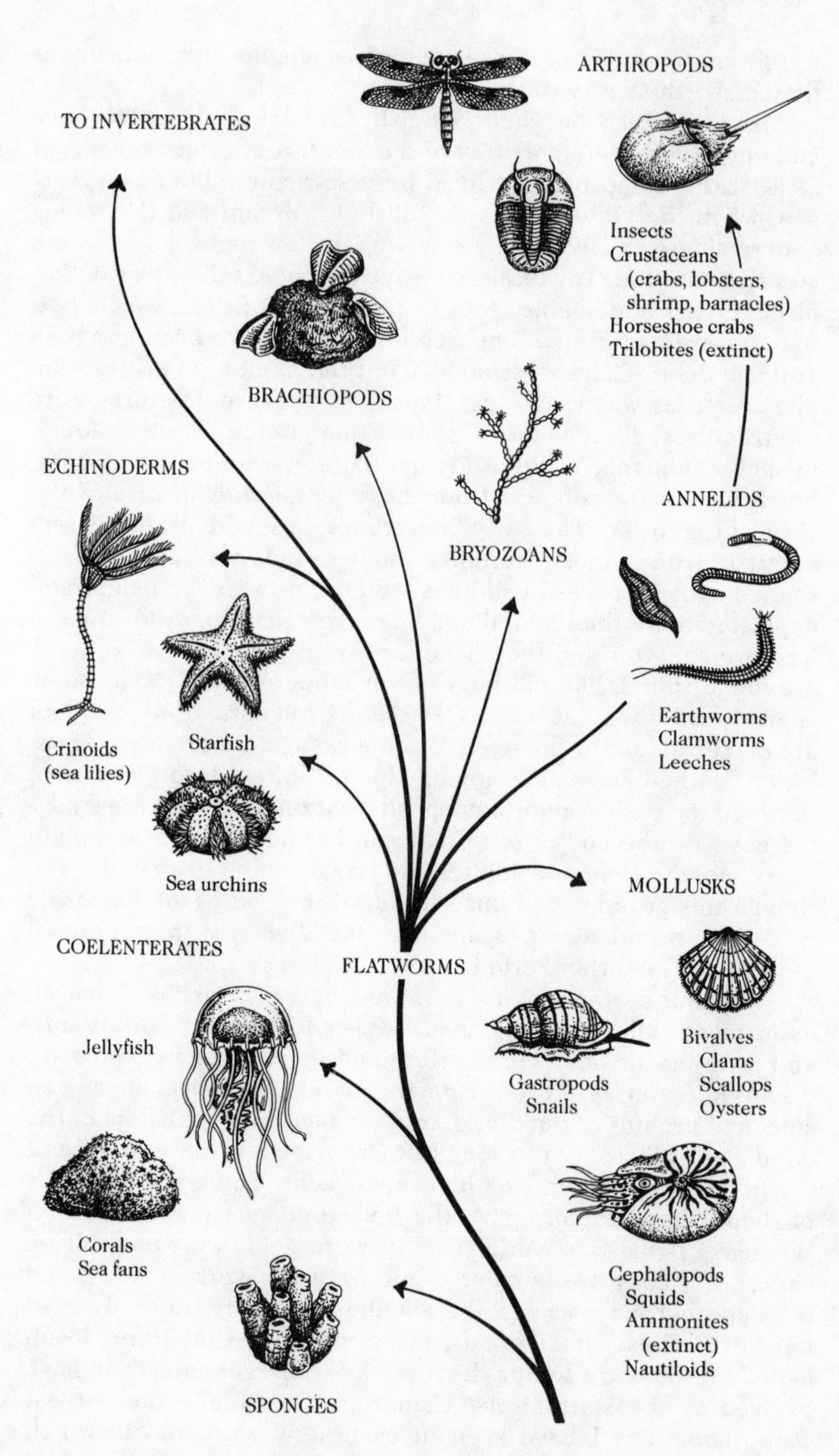

Evolutionary pathways leading to major groups of animals.

shelled fossils that tell of unknowable numbers of creatures, brachiopods in the trillions, lying on the muddy sediment at the bottom of a quiet, warm, shallow sea. They are the most characteristic element of the Paleozoic Era. They shared the seas with many other hoary creatures, such as the trilobites and crinoids; but it is the brachiopods that typified the first great era of skeletonized life. What happened to them? Animals so well adapted for a sedentary life of filtering the rich plankton, a lifestyle similar to that of modern clams—why do so few of them survive today?

The Rise of the Brachiopods

In the early 1970s I took a course on comparative invertebrate zoology taught by two great zoologists, Paul Illg and Alan Kohn, of the University of Washington. For over twenty weeks we studied the various invertebrate fauna, spending long, demanding hours in examination and dissection, as phylum by phylum we surveyed the various major groups of the world's animals. I was fascinated by the biological diversity of the world, but my devotion to the various phyla was not uniform: I was most interested in those groups of creatures that have materially contributed to the fossil record. It was therefore quite a disappointment when brachiopods were dismissed after a single lecture and lab period. I was at first incensed; how could such a noble group be given such short shrift? But my sensibilities had been skewed by the nature of the fossil record, which is largely the record of those creatures whose remains have been preserved—in the sea, creatures with calcareous skeletons—and in the modern world, and probably in the past as well, far more creatures have existed without skeletons than with them. The truth of the matter is that brachiopods are but a very minor component of current ocean ecosystems, clinging and hiding in a few relict habitats scattered around the world, pathetic shadows of their dominant Paleozoic ancestors. More than 3000 genera of brachiopods and tens of thousands of species are known from the past. Barely 300 genera still exist today.

Brachiopods are subdivisible into two major groupings: those with hinge mechanisms on their shells and those without. The latter group, known as inarticulate brachiopods, were the first to appear. At the Addy quartzite in eastern Washington and at countless other Lower Cambrian localities throughout the world, we find these tiny shells in association with the first trilobite fossils. After several millions of years, the inarticulate brachiopods were joined by the articulates, which evolved a better system of holding their two shells together: they had calcareous teeth and sockets very similar to those of clams, and they served the same purpose. When the two shells are held together in this way, they can open and close with greatly enhanced efficiency. This sytem also lessens the number of muscles necessary for opening and closing, and thus frees space within the shell for other anatomical features, including an enlarged feeding apparatus, the lophophore.

At first only a few brachiopods used this more efficient, articulated shell system. The inarticulate forms held sway, living in vast numbers in early Cambrian ocean habitats. But gradually the balance of the two groups began to shift. In head-to-head competition for food and in ability to survive attack by predators, the articulate shell design must have proved superior, for the numbers of the articulate brachiopods continued to swell while the numbers of inarticulate forms declined. By the end of the Cambrian Period, some 500 million years ago, the relative abundance of the two groups had been reversed; the number of inarticulate taxa had dwindled to almost nothing, while the articulates continued to diversify into thousands of species and untold numbers of individuals. By the time the early Ordovician strata abundant in the Ohio Valley were deposited, they were perhaps the most numerous creatures in the seas, living in a wide range of habitats. Some attached themselves to rocks while others lay flat on sand or mud; they lived in the shallows and at great depths, and in all the waters in between. From about the middle of the Cambrian Period until the end of the Paleozoic Era, a span of about 250 million years, the articulate brachiopods were one of the most successful groups in the sea.

Evolution is a bit like life insurance; individuals live and die in unpredictable ways, but the trends of larger numbers are both observable and predictable. Which group survives and which dies is simply a numbers game, usually the result of only the slightest changes in survival frequencies. The tiniest advantage offered by a slightly different way of living, or a new morphology for feeding for instance, will be magnified over the millions of years and generations of evolutionary time. So it was with the brachiopods. During the Cambrian Period the articulate forms must have had some slight advantage over the inarticulate forms. For reasons related to feeding efficiency, perhaps, or an increased ability to withstand attack by hungry predators because of the strength of a hinged shell, the articulate brachiopods superseded their earlier-appearing cousins in the Cambrian sea bottoms. More often than not, such slight advantages will lead to the complete extinction of the inferior group. The inarticulate brachiopods nearly suffered this fate. Nearly, but not quite.

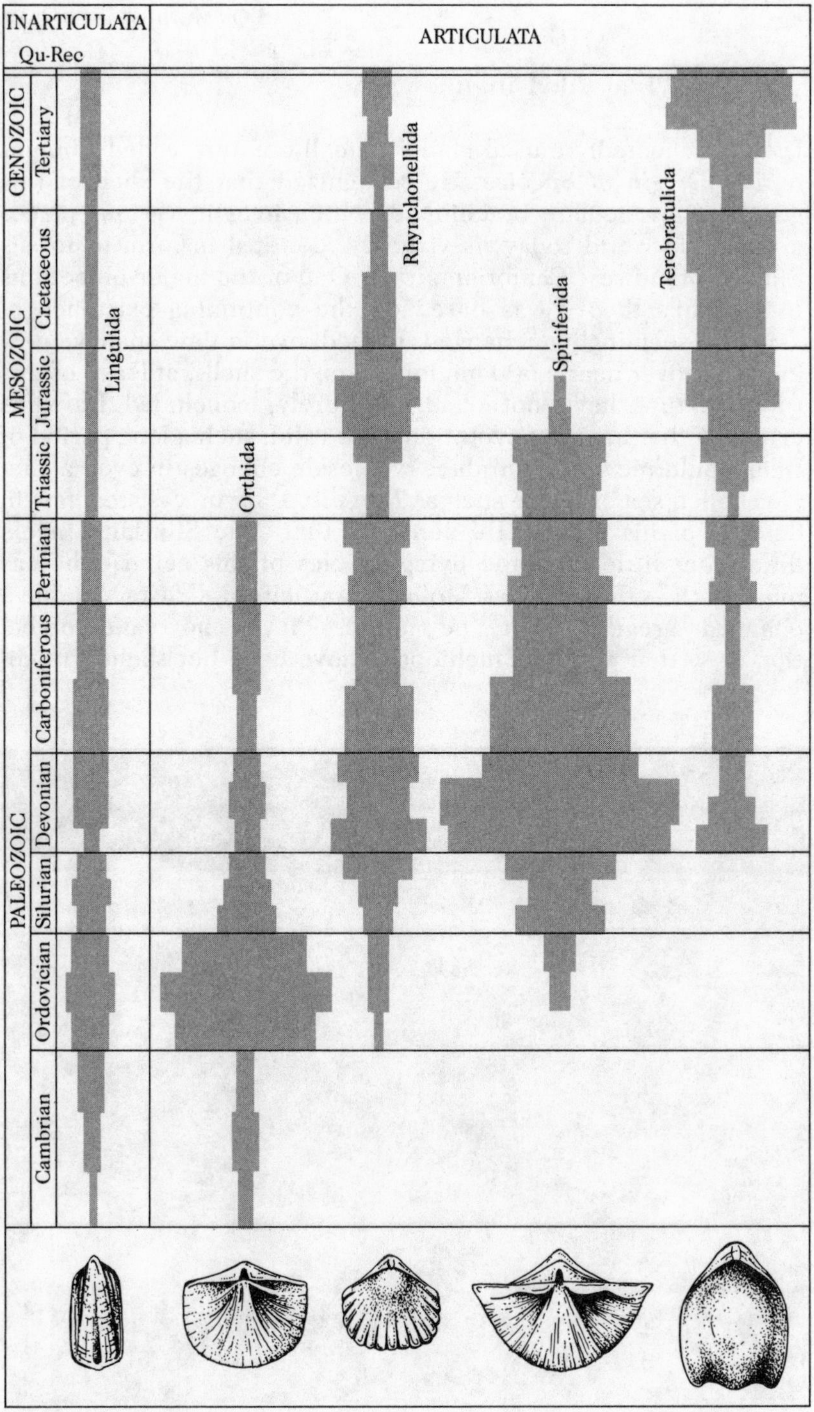

Geological ranges of the brachiopod orders. Lingula, *an inarticulate brachiopod is shown at the left, the articulate brachiopods to the right. The widths of the boxes indicate the proliferation and shrinkage of the brachiopod taxa of the various orders during the time periods indicated.*

The Survival of Lingula

Darwin repeatedly mused about a small creature called *Lingula* in *The Origin of Species*. He recognized that the shell of this genus of inarticulate brachiopod, which lives in various places around the world today, is virtually identical in form to fossils found from lowest Cambrian strata. One of the major objections to Darwin's theory was based on the continuing existence of *Lingula*, seemingly unchanged. Indeed, over a time span we now know to cover nearly 600 million years, the shells, at least, of this tiny creature have not changed. Surely, concluded Darwin's critics, if the theory of evolution were valid, such a long period of time would inevitably produce wholesale changes in even a simple shelled sea creature such as *Lingula*. Darwin was steadfast in defense of his theory. He admitted that "the Silurian Lingula differs but little from the living species of this genus; whereas most of the other Silurian Mollusks and all the Crustacean have changed greatly." But, he noted, "it is no valid objection . . . that certain Brachiopods have been but slightly modi-

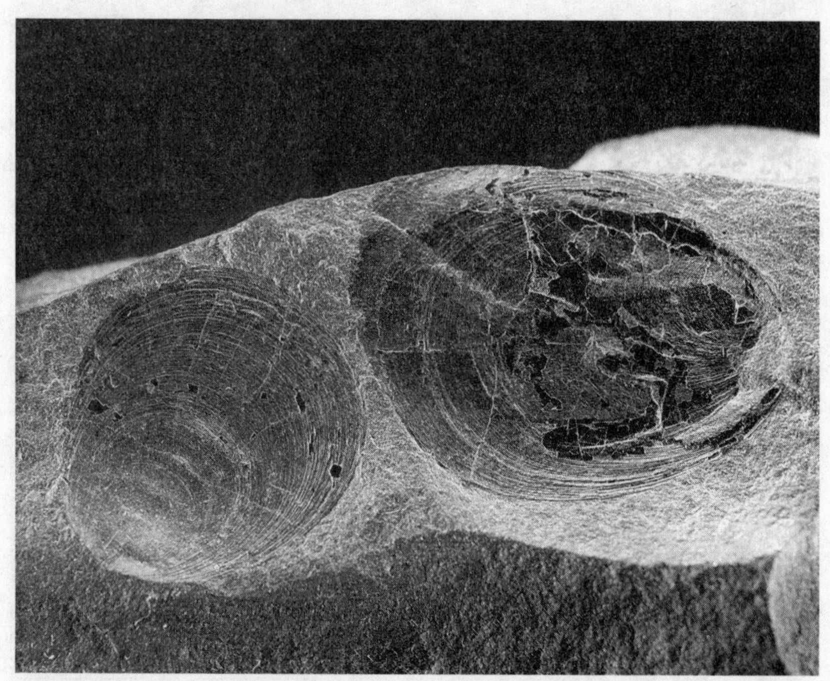

Lingula, *an inarticulate brachiopod, a form of life that has persisted from the Ordovician to the present. (Specimen from North Museum, Franklin and Marshall College. Used with permission of Grant Heilman Photography; photo by Runk/Shoenberger.)*

fied from an extremely remote geological epoch. When advanced to a certain point, there is no necessity, on the theory of natural selection, for their continued progress; though they will, during each successive age, have to be slightly modified, so as to hold their places in relation to slight changes in their conditions."[3]

If a creature was to survive for long periods of time, it has either to adapt or to avoid the "slight changes in conditions" which Darwin considered to be the fuel of evolutionary change— and of extinction. But changes as organisms perceive them can come in many guises. The most obvious are physical changes— differences in such basic aspects of the environment as temperature, oxygen level, the salinity of the sea. But what about changes of a more biological nature? In the 600 million years since the first appearance of creatures that looked like the modern-day *Lingula*, virtually every aspect of marine ecosystems has changed enormously. Almost every creature of the Cambrian, if it were somehow brought back to life and released into our seas, would meet a swift end, usually in the mouths of the very able and efficient predators that have evolved in the intervening 600 million years. How has *Lingula* managed to survive for so long? The answer seems clear. *Lingula* was one of the first creatures ever evolved to burrow. But many creatures burrow today. *Lingula* made an even more fundamental move that aided its survivability: it learned to withstand waters of lowered salinity, and thus it can live in environments intolerable to most competitors and predators. Sometime early in the Paleozoic Era some small stock of lingulids evolved the ability to live on the brackish margins of the sea, in such places as high-runoff seashores and the edges of river mouths and estuaries. Articulate brachiopods need completely normal marine salinity to survive. The lingulids found themselves in a place with neither competitors nor shell-breaking predators. The other inarticulate brachiopods, those that continued to live in the open sea alongside the articulated forms, were nearly all gone by the end of the Cambrian Period.

I have seen lingulids alive only once. I found them on a sandy beach in Fiji, quite by accident. I was digging in the sand, looking for mollusks, and found only these bivalved creatures, seemingly clams at first glance, but certainly not. The emptiness of the sand was evidence of the harshness of this environment, for the rich bounty of the tropics will fill every environment with creatures, given the chance. But this sandy beach had lingulids and little else. I marveled at their long-ago ingenuity in evolving the ability to live where virtually no other creature can. To me they are a

[3] Darwin, *Origin of Species*, p. 271.

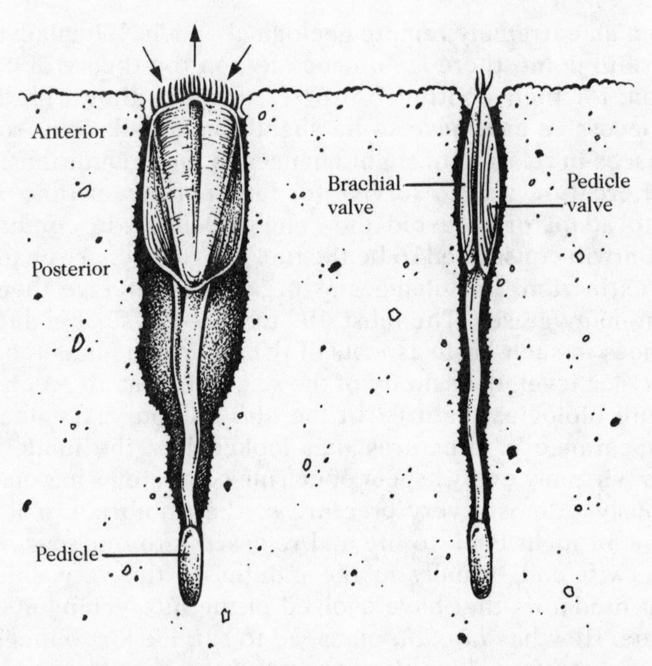

Typical shell morphology of Lingula, *shown in the living position.*

triumph of Darwinism, not a challenge to it. They are one of the oldest creatures still living on earth.

The Greatest Extinction

As you walk the land, bone-dry land, it takes a long reach of the imagination to believe that a shallow sea once existed here. And yet such was the case in this West Texas desert. You are surrounded by sagebrush and tumbleweed, and although it is still early spring, the sun already cooks the dusty terrain. You are hiking up through a dry wash, crossing a wide alluvial fan deposited by Bone Springs. Low rocky outcroppings can be seen crumbling, but the fine shales are too weathered to yield much insight into the ancient sea of their origin. As the morning wears on you climb higher still, passing upward into the multicolored rocks of Cherry canyon and finally up into Brushy Canyon. At these higher elevations you find more consolidated rock, and sea fossils galore: mostly brachiopods, brachiopods everywhere, but species very different from those of the Ordovician-aged Ohio Valley. You are high above the desert floor now, and the West Texas winds blow ceaselessly. Buzzards wheel in giant circles over your head, hoping you will join the long-dead brachiopods in the next world perhaps. But your attention is entirely on the giant cliff of searing white rocks thrusting upward above the soft shales you stand on, white limestones of Late Permian age. You finally reach the base of these white cliffs and begin the demolition for which you have come. With hammer and chisel you break off large hunks of white limestone and peer with wonder at an abundance of fossils, packed into every nook of the rock, a carnival of long-extinct creatures. You see beadlike fossils of a type completely foreign to you, the remains of calcareous sponges once common. Bits of calcareous algae and bryozoans are also evident, along with fragments of long-extinct corals. And interdispersed with this richness you find uncountable brachiopod shells. You are collecting El Capitan, a giant block of limestone extending from West Texas to New Mexico, the remains of a giant barrier reef complex that made up the southern coastline of North America about 250 million years ago. You are seeing the remains of what was perhaps the largest single reef system ever to exist, a structure that would have dwarfed even the present-day Great Barrier Reef system of Australia. It is a sad monument in a way, for this great, ancient burial ground records the last flowering of Paleozoic life before the Fall—the single greatest mass extinction recorded in rock history, which occurred at the end of the Permian Period. Within several millions of years after the rocks around you were deposited, most of these species became extinct, along with as many as 95 percent of all other species on earth. The Fall ended the Paleozoic Era, and with it the hegemony of the brachiopods.

It is somewhat ironic that this greatest of known extinctions has received so little attention in relation to other such events in the earth's past, particularly the extinction at the end of the Cretaceous Period, 65 million years ago, which did in the dinosaurs and much else. Perhaps this neglect can be attributed to the much greater age of the Permian event, for it occurred almost 250 million years ago. Or perhaps it came about because the creatures it affected are so little known in comparison with the dinosaurs. But let no one be fooled—the extinctions that closed out the Mesozoic Era were but a shadow of the grim reaping that occurred at the end of the Paleozoic. At the end of the Permian the face of death was to be seen everywhere, both on land and in the sea. Only one of every ten species survived the end of the Permian.

What caused this mass dying? There is no evidence of great volcanic paroxysms, or of giant meteors flaming through the atmosphere to strike the earth with deadly force. Earth scientists familiar with this extinction don't even believe that it occurred rapidly; most of them suspect that the event lasted longer than a million years, and may have gone on for more than 10 million years. This great death seems to have been triggered by no extraterrestrial event, such as the crashing to earth of a meteor or comet, but by events caused by the changing face of the earth itself. The agents of death at the end of the Permian Period appear to have been a change in climate and a lowering of the sea level, both created in large part by the positions of the continents.

Geologists would very much like to know what coordinates or controls the movement of the continents. The great revolutionary theory of plate tectonics, formulated in the early 1960s, has shown that the sea floor spreads and huge regions of the earth's surface drift a few inches each year. These huge plates, some carrying continents, some not, move about over the surface of the globe like children in bumper cars.

At the end of the Paleozoic Era all of the continental masses we know today coalesced into one huge supercontinent. For the only time that we know of (no such thing has happened before or since), all of the major continental blocks lay welded together.

It took tens of millions of years for the various continents to converge in this supercontinent, which has been named Pangea. As the various land masses came together, the climate of the world changed drastically. The interiors of continents in our world are mainly places of climate extremes—hot summers and cold, harsh winters. As the continents merged, the interior areas cut off from the moderating influence of the seas became ever larger. These continental interiors must have been among the least hospitable places in all the long history of our earth. The strata found from these environments tell a tale of drifting sand and salt deposits, stark testimony of aridity. And as the world climate changed during this process, giant ice sheets began to grow over both the north and south polar regions. One of the greatest glaciations in the history of the earth unfolded. As the great ice sheets advanced, sucking up moisture from the air and sea, the level of the seas began to drop, rapidly draining the most favored of marine habitats, the shallow shelf seas of the world, where nutrients and light are so abundant. No wonder so many species on land and sea began to die. By late-Permian time only the tropics maintained a preserve of abundant animal and plant life. The giant Permian reef complex of West Texas is a last bastion of life in this long-ago late-Paleozoic world. And then that life, too, slowly died.

The list of victims of the late-Permian extinctions is long. Prominent among them were the trilobites and all of the Paleozoic corals, most crinoids, and large numbers of land reptiles. Among the hardest hit of all were the brachiopods. *Lingula* survived, as did a handful of articulate brachiopods. But many thousands of species of brachiopods did not.

With the start of the Mesozoic Era the few survivors faced an emptied world. The continents began to spread apart and the glaciers retreated; the sea rose and flooded the continents. The huge inland seas rich in nutrients opened up new opportunities for the rapid evolution of sea life. New species of brachiopods began to form. But history did not repeat itself: the ocean habitats did not fill with a diversity of brachiopods, as in the Paleozoic, for the newly evolving brachiopods found their old haunts already occupied; "no vacancy" signs were already hung out over the offshore, normal marine salinity environments so beloved of the Paleozoic brachiopods. The new tenants, bivalve mollusks, were so well ensconced that the brachiopods were unable to regain a toehold in their marine realm. The new masters controlled the feast of plankton; the old were sent to the ecological sidelines to eke out marginal existences in caves and deep water or, like *Lingula*, at the edge of the sea. The heyday of brachiopods was over, never to return.

Evolution is a numbers game. More bivalves than brachiopods survived the Permian extinction, so the bivalves had a head-start in repopulating the early-Mesozoic world. To make themselves unappetizing to predators, the surviving brachiopods became poisonous to eat. To find a home of their own they retreated to inhospitable habitats and strong-current areas of cold-water oceans; except for a furtive existence deep in the caves of reefs, they quit the tropics entirely. To see brachiopods today you have to be ready to dive deep into cold water.

A Day in the Life

Some days are made for diving in Puget Sound. On cold winter days you have to be crazy to put on a thick wet suit and descend into the frigid water of the fjordlike waterway of northern Washington State; in fact, with an average water temperature of about 45°F, many people think you would have to be crazy to dive in there at *any* time. But on some days the air is so clear and the sun reflecting off the mirrorlike surface of the green water so warm that a dive into the rich waters of the sound seems like a perfectly sane idea.

I am mulling over such earthshaking thoughts as I pull on my long johns. I am sitting in a friend's boat, an old cabin cruiser, on a perfect July day in 1980. We are adrift off Vashon Island, a small isle about thirty miles south of Seattle. My diving buddy is an old friend, a man I have dived with for a decade. The diving fraternity in the Seattle area has been a small group for many years, although that situation, like so many others, is changing these days. We taught scuba diving classes together when I was an undergraduate in college, so there is no need on this or any other day to go through the macho exercises so beloved of many divers of our acquaintance. Which means that we are both bitching like mad about the fact that the wet suits are wet and must have shrunk (it couldn't be that our waistlines have expanded, after all), that the tanks didn't get an honest fill, that the regulators aren't breathing properly, and so on. This is ritual; we wouldn't feel good about going diving if we didn't go through this exercise. We are nearly ready now; both of us have put on our bubble suits and have swung the massive tanks on our backs. Just before rolling over the side of the boat, my friend turns to me and asks, "Hey, Doc, just what the hell *is* a brachiopod, anyway?" I give him my best raised eyebrow and fall back over the side of the boat into one of the few places on earth where these survivors from the Paleozoic Era still flourish.

I began diving when I was sixteen. I bought a fire-extinguisher bottle and a new valve for it, and a friend gave me an old double-hose regulator. I got an air fill, a mask, and a pair of fins. Thus equipped, I dived into Lake Washington alone. I will never forget the exhilaration that comes when you realize that you do *not* have to go back up after a few seconds, the sense of freedom when you take your first breath off an aqualung underwater. But it is very easy to die in the water, diving alone, not knowing or caring about embolisms or pulmonary emphysema or a score of other horrible fates. I can still scare myself silly just thinking about it. God sometimes looks after his more stupid children, and I survived to find new equipment eventually and to read some books. The equipment is a lot different now; the air suits keep you much warmer and the regulators are much more efficient. But I experienced my greatest joys with the cadged, jury-rigged outfit of my teen years as I learned the waters of Puget Sound, developing advanced skin wrinkling before my time.

No matter how good the fit of your suit, with its hood and gloves and booties, there is always some spot where at least a little cold salt water finds its way to your skin when you first hit the water. On this day I find, to my shock, that the small hole in my wet suit, very inconveniently located right above the small of my back, has not been repaired. I reorient myself after the somersaulting entry from the boat, find "up," and look for my friend. I see him in the distance. We are fortunate on this day, for the water visibility is about twenty feet—a very clear day for Puget

Sound. Some days you're lucky to see your hand in front of your face. I swim over to him with good hard strokes of my fins, and hard strokes are needed, for the tidal current is still running, though rapidly diminishing, and it takes most of my strength to push upstream.

We have chosen this day because it promises little tidal change. Puget Sound is mesotidal, which means that the tides can rise or fall as much as fifteen feet in six hours. Such a rapid change in the height of the sea in a restricted body of water produces swift and powerful currents. These tidal currents are one of the main reasons that Puget Sound teems with one of the most diverse assemblages of marine creatures known on earth.

A diver can swim, at best, about one-half knot. We are both slightly heavy, and float downward together along the anchor chain, finally reaching the sandy bottom in about 20 feet of water. The bottom here is slightly rippled and pocked with rocks. The larger rocks, some as big as a football but most the size of a fist, are all covered with barnacles and mussels, and here and there the starfish that feed on them. Bright-red rock crabs scuttle among the rocks and raise menacing claws at our passage. All of the sediment here is the refuse of glaciers, the remains of the scour and gouge of the monstrously thick piles of ice so recently part of the Northwest landscape. Most of Puget Sound is bordered by high cliffs composed of sand and gravel; heavy rainfall ensures that large quantities of this material make their way into Puget Sound.

We have chosen this spot to dive because it is one of the few areas in the southern part of Puget Sound where vertical cliffs are exposed underwater; in most areas the bottom slopes downward gradually. We swim along the bottom, edging downward in colder water. Warm summer is only about 20 feet above our heads, but the salt water rapidly darkens as we move into deeper water. I can feel the increasing pressure reducing the volume of my air suit until the fabric is tightly clutching my skin, the giant invisible hand of Boyles Law reminding me of my journey into an evermore foreign land. I fumble for the valve to admit more air into my suit, and the rush of warm air is a relief. I glance at my partner, moving gracefully and silently at my side, and I am heartened by his presence. I am but a guest in this dark world, if a frequent one; no matter how often one dives into this body of water, there is always the slight sense of dread with the descent into the dark and the cold.

As we move downward over the slope we see the sediment change, and the animals too. The cobbled, rippled sand gives way to a finer, purer sand, and then to silt. At 30 feet we enter a magical realm, a forest of bright-yellow sea pens. These creatures, related to soft corals, look like plants, but they are highly integrated colonies of animals. They are about two feet high and in some places are no more than a foot apart. They endlessly filter the surrounding seawater of its cargo of plankton in this frigid place, and are themselves the food of starfish, which slowly move among them. The sea pens may be living fossils, for fossils that look much like the creatures of Puget Sound are known from the Precambrian-aged Ediacaran sandstones of Australia. Once in a while we scare a flounder off the bottom, and as we pass a sunken log a large octopus disappears into the rotted wood, the site of its den betrayed by a pile of crustacean refuse. Finally, at a depth of about 50 feet, we come to the edge of a great underwater cliff.

I float over the edge, weightless, and ready the camera I have brought along. My underwater Nikonos has been a reliable friend for many years and has witnessed memorable scenes. I screw in the appropriate settings in the dim light and ready the attached strobe, for the available light down here is far too dim for picture taking. My friend has a large underwater light, now on, and moves over the side with me. We let a little air out of our suits to reduce our buoyancy and float down toward the absolute blackness beneath us. At about 75 feet I move in close against the wall and see the creatures I have come to collect.

The sheer wall here is another legacy of the glaciers, and of the tidal currents as well. It is composed of cobbles and boulders of Pleistocene glacier outwash, sediments shaped and eroded first by the glaciers and then by the tidal currents that wash back and forth four times a day. I look closely at this underwater wall and see countless brachiopods, each shell gaping slightly, drawing in seawater from each side and pouring it out of the front. The brachs are at most about an inch long, and in the glare of my friend's light I see dull brown shells with bright-orange interiors. Each individual brachiopod is strongly anchored to the wall with a thin proteinaceous tether, called a pedicle—its lifeline. If ever this cord is broken, the brachiopod cannot reattach itself, and will die. The cord is tough, and the cement attaching it to the rock surface is tougher yet. Many scientists and chemists have pondered the chemistry of this glue, so strong that it makes our synthetic cements seem ludicrous by comparison. I set up a photo at close range; the ensuing strobe blasts the surrounding area into whiteness for the briefest of moments. I wonder if these eyeless shellfish can detect this sudden release of energy on any level at all, and the thought leaves me cold. Our lives are controlled largely by light, and we define our world mostly in terms of its visual context; it seems (from my anthropocentric viewpoint) so alien to confront a creature that has never made the slightest evolutionary accommodation to light. As my eyes readjust to the darkness I look again at the pale shells on the wall before me. They grow slowly, these brachiopods, taking five to seven years to reach their full length of about an inch. Almost everything we know about their natural history has come from the work of one man, Charles Thayer of the University of Pennsylvania, and I wish Charlie were here now to explain just what it is I am seeing on this deep wall. Is this an ancient brachiopod population? A new one? What is its past? And what of its future? I have long promised this dive to Charlie, to show him this wall.

We have drifted ever deeper, for I want to map the face of the wall and learn the depths at which the brachiopods drop out. But they stay present in undiminishing numbers as we drop below 100 feet. It's very dark now and very cold; the great pressure requires us to admit much more air into our dry suits to maintain neutral buoyancy. Air is now a consideration, for each breath we take pulls a large volume of compressed air from our tanks; at this depth we're going through air at a rapid clip. We also must worry a bit about the nitrogen uptake of our bodies; we have only twenty minutes at most. Any longer and we'll have to decompress on our way up if we're to avoid the bends.

I can keep track of my friend because of his light. Even so, we must take great care to stay together, for the visibility is all but nil. Signals of many kinds tell me it's time to go home: my watch tells me how long I've been down here; the pressure gauge on my tank tells me of the relentless dwindling of the life-giving air on my back; my body temperature, despite the cumbersome dry suit, is rapidly dropping, and I'm beginning to shake. Our depth has also brought a slight edge of nitrogen narcosis, leading to a foreboding that cannot be erased. But mostly I'm concerned about the tidal current. For as long as we have been down here, more than fifteen minutes now, the water has been virtually still: we planned this dive for the time of slack water, the cusp between the tides, when the great volumes of seawater are held in the balance. But now I begin to feel the tug of the outgoing tide start to push against me. We must leave now if we are to get back to our boat. I move back to the wall and with my knife dislodge several brachiopods to take back to the aquarium at my university. I place them in my goody bag and turn to look for my friend. With shock I realize that he dropped down beneath me; I can't see him, but the stream of bubbles coming up around me indicates his position. I hear the twang of his spear gun firing. The noise, as usual, gives no hint of direction. I drop down through the stream of his rising bubbles and find him struggling with an enormous ling cod, the top carnivore of this world. The fish is speared but not dead, and it struggles mightily. There is little I can do but watch as my friend finally subdues the fish and attaches it to a stringer dangling from his weight belt. I give him a vigorous thumbs-up sign. He returns a hearty nod and we begin our ascent into the light. But the current is pushing with a vengeance now, and as we rise we are pulled northward. It's useless to try to swim back against the current; that mistake is often fatal to divers in this area. Instead we move up the wall until we reach the lip and then strike toward shore, now being carried swiftly by the relentless current. It is this current that brought the brachiopods to this place, this swift river of seawater rich in suspended organic material and pastures of plankton, a movable feast perfect for a tiny creature firmly anchored to the substrate.

With relief we finally reach the 20-foot depth once again, with its warmth and light. We ascend the last feet and break surface to see our boat several hundred yards to the south. We swim to the beach, kicking hard now, perpendicular to the direction of the current. We are tired as we flop on the shore, but before trudging down the beach and then swimming out to the boat we inspect our respective catches. I make appropriate comments about the puniness of the ling cod. My friend takes this ribbing gravely and makes a point of inspecting a brachiopod from my goody bag and dismisses it as "just a dumb clam." Then I launch into my sermon about the differences between clams and brachiopods until a wicked smile lets me know I've been caught pontificating again. "Do you mind?" my friend asks, and without waiting for a response, he wedges the end of his knife between the two shells, now firmly closed. I wince at the destruction of this prize, but he finally pops the shells open, ripping several shell-closing muscles in the process. My friend looks at the opened brachiopod with surprise. "There's nothing in here," he says, but I point out the lacy, frill-like lophophore, the specialized feeding structure that makes up the greater part of the internal organs of a brachiopod. My friend has eaten many a clam in his life, and even if he is completely ignorant of clam anatomy, he quickly realizes how different the internal anatomies of a clam and a brachiopod really are. Most people are fooled because of the similar shape and size of the shells, but once past the shell, you are obviously dealing with a very different animal.

My friend holds up the remains of the now thoroughly destroyed brachiopod and has derisive things to say about how little flesh there is for so large a shell. "No wonder they're almost extinct," he mumbles, and then asks me if they are any good to eat. "Don't do it!" I tell him. My friend assures me that *anything* taken from Puget Sound waters is edible. *"Don't do it!"* I shout. He gives me his best stage sneer, the one reserved for scientists, and intones his favorite condemnation of scientists and their practical ignorance: "Too much college and not enough high school." The brachiopod, remnant of one of the great stocks of life, survivor since the earliest Paleozoic, disappears into his mouth and plays its part in the ongoing process of evolution. It's death surely serves some purpose, for it greatly diminishes the probability that at least one human being will ever again try to eat a brachiopod: even before the wretched creature is halfway down, my friend turns green, and retches violently on the beach. I can't help myself; I roar with laughter. "One thing about college," I tell him when he regains his composure and control of his stomach. "At least they teach you not to eat brachiopods."

Paleo

The other brachiopods I gathered from the wall on Vashon Island on that warm summer's day are now spread out on aluminum dishes, torn and mangled. They have been dissected by my invertebrate paleontology class. Week by week over the ten-week term at my university we have been surveying the major groups of creatures that have contributed to the fossil record. Each year I teach this class; it becomes one of the markers of time in my life. It is late afternoon and already dark outside; the winter days in Seattle are short and bitter. As always after the three hours of lecture and lab, I am drained and contemplative as I survey the resulting carnage. Next to the dissecting dishes on the lab benches are specimen drawers, each holding many hundreds of fossil brachiopods for the students to examine. The science of geology has advanced rapidly over the decades, and many of our undergraduate classes are conducted in computer rooms and mass spectroscope laboratories. But paleontology still involves the handling of countless specimens and the rote memorization of their names. The process seems archaic to the students, hopelessly old-fashioned. I sympathize with their complaints about the amount of information they are required to learn. In paleontology there is no easy way, no shortcut. The best I can do is show the students that the field still fascinates me.

I gaze around the room at the brachiopod collection; it is huge and magnificent, easily one of the best teaching collections in the country. My predecessor and first professor, the great V. Standish Mallory, assembled this collection over his long career. Two decades ago I was his laboratory instructor for this same class, and now, irony of ironies, I am back teaching this laboratory once again.

I manhandle the cumbersome drawers back to their squat cabinets: spiriferids, orthids, pentamerids, names now like old friends; lingulids, productids, atrypids pass before me like fallen armies. The brachiopods make up the bulk of the collection; they are present in unbelievable diversity. Yet in drawer after drawer the ages of the various specimens are the same: Paleozoic Era, Paleozoic Era, Paleozoic Era; only occasionally does a Mesozoic or (rarer yet) a Cenozoic specimen sneak in. This room is itself a crude record of the waxing and waning of evolution: the Paleozoic heyday of the brachiopods and their Mesozoic and Cenozoic decline are adequately documented in the number of drawers alone.

With today's lab we have finished the brachiopods. Their meteoric rise in the Cambrian, success in the Paleozoic, and virtual extinction at the end of that era have been documented. I start pulling out the drawers of the next group to be studied in my class, the bivalve mollusks. They are much like the brachiopods in morphology and lifestyle—creatures with two shells but no head, passive filterers of plankton—but they differ considerably in their distribution through the ages. My collection of early-Paleozoic bivalves is small—only a handful of specimens. I have more from the late Paleozoic, but the collection is still small in comparison with the cornucopia of brachiopods known from that time. In the Mesozoic, however, my collection of bivalves skyrockets, and it increases even further in the Cenozoic. By the time I have finished, the tables are once again nearly covered.

My long day is over. I once more look over the specimen-covered tables as I leave. The specimens are arranged in the order of the strata in which they were found—first the oldest clams, then table by table, younger and younger ones. Amid the diversity of form a rough order is discernible. In the middle of the room, on the tables reserved for the Mesozoic-aged collections, I see forms virtually absent from the younger specimens. I suddenly look forward to the next lab and the story of the flat clams, creatures that, like the brachiopods, have been bypassed by the history of life.

3

Before Modern Predation
The Flat Clams

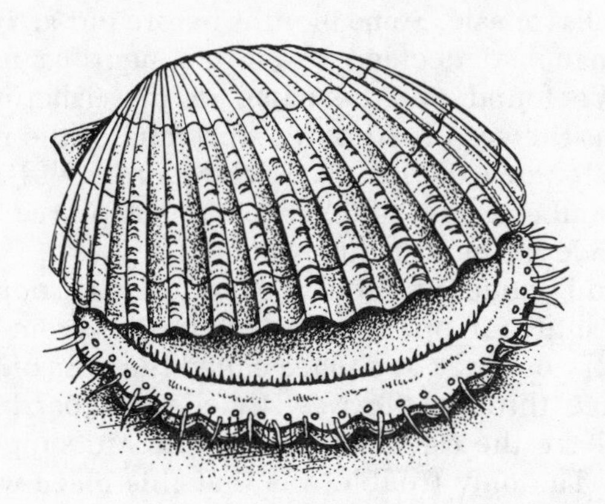

Inaccessible Bay

During the austral winter of 1981 I was engaged in studies on the biology of the nautilus, the last living shelled cephalopod. I was working with a colleague, Lewis Greenwald of the Ohio State University, and we had decided to conduct our studies in one of the world's most beautiful places, the island of New Caledonia, located about 700 miles to the east of Australia's Great Barrier Reef. Lew and I had arrived with an impressive list of experiments that we intended to carry out, but as the weeks piled up and the specimens did not, that list came to seem increasingly like a fantasy. This was my fourth visit to New Caledonia, and until now I had never had any difficulty in getting a steady supply of specimens for my experiments. Nautiluses are scavengers, and they can detect carrion in the water at great distances. By building large wire cages, baiting them with meat of some sort, and placing them at the front of coral reefs at depths of about 1000 feet, I could quickly capture any number of nautiluses. But on this trip our luck was atrocious. We'd be up early and head out to sea in our small boat before the trade winds began to rise; we tried to have our traps in the water soon after dawn. We attached the traps to surface buoys with long polypropylene lines and left them there. Two days later we'd go out again, find our buoys, and winch up the traps, from their thousand foot deep resting places. But time after time, the long winching process would be culminated in an empty trap.

The seeming disappearance of nautiluses from an area where they had once been fairly abundant was inexplicable at first, but we soon discovered the reason. In every curio shop around the town of Nouméa, the capital of New Caledonia, we saw numerous nautilus shells for sale. Some months before our arrival a commercial fisherman had decided to start a nautilus fishery around Nouméa. We found out the name of the fisherman and soon located him. He was proud of his work and of the money he was making. He let me examine his catch records. With a heavy heart I saw his numbers: over 3000 specimens captured and killed for the shell trade in the last three months.

Lew and I pondered a course of action, and soon decided that we had to shift our operations ever farther from Nouméa. We decided to try our luck at a place called the Pass of St. Vincent, a large passage through the New Caledonian barrier reef which seemed to have the right bottom contours to support a nautilus population. The only trouble was that this place was more than thirty miles to the north of Nouméa, and our boat was only 19 feet long. We watched the weather patterns, and when the sea promised to stay calm for a few days we decided to try our luck on this long voyage.

We set out in the early hours of a glorious day and headed north along the coast. Here the barrier reef was over ten miles from shore, and to lessen the effects of wave action on our boat (and backs) we hugged the coastline. It was a dreamlike voyage. We passed jungled river valleys and savannahlike coastal terraces, steering around the many shallow patch reefs marked on our charts. For three hours we continued in this fashion, in the best of spirits, before finally sighting our pass.

The barrier reef surrounding New Caledonia is second only to the Great Barrier Reef of Australia in size. It parallels the coast for 800 miles, broken only occasionally by a pass to the open ocean beyond. We hoped that the pass called St. Vincent was far enough from Nouméa to have escaped the ravages of the shell dealer.

With some trepidation we left the coastal waters and headed out to sea. The reef was marked by giant breakers, for even

Inaccessible Bay, New Caledonia.

though we were in the friendly tropical latitudes, the relentless storms of more southerly regions filled the oceans with giant winter waves, some of which found their way northward to smash against the reefs of New Caledonia. The pass was about a half mile wide. Entering it was like entering a mad roller coaster, for the huge rollers coming in from the ocean became compressed and focused as they entered the pass. The pass was far too deep to allow the waves to crest and break, but it did produce a tumultuous sea.

Over the next several hours we baited our traps, attached lines, and lowered this ponderous gear over the side of the boat, keeping an eye on the echo finder to make sure the traps would end up on the 1000-foot bottoms that promised to hold the largest concentration of nautiluses. When three traps were finally set, their buoys riding the long roller-coaster swell, we were ready to head for home.

We faced a long voyage back to Nouméa against a stiff wind. By this time Lew and I were exhausted from the trip, the excitement, and the hot sun; and especially from the hard, exacting work of setting our traps in such depths. We decided to make for the nearest land and camp for the night. At this latitude we had no need for blankets; mosquitoes, not cold, would be the greatest impediment to a sound night's sleep. Far out to sea, twelve miles from land, we scrutinized our chart. It was now late afternoon; we had only an hour of daylight left to find our way ashore. We chose a likely looking bay and headed for it.

Although I had prepared Lew for a night of tropical camping, it occurred to me that we might manage a more comfortable night's sleep. On my last visit to New Caledonia, a year before, I had befriended a man who had recently arrived from Marseilles. His family had long been engaged in the commercial raising of oysters in the Mediterranean, and now he had come halfway around the world because he had a novel idea: the French have a high regard for oysters, and Maurice was sure he could make a fortune by growing and selling the mollusks to the large French population of New Caledonia. He had bought a large stretch of deserted coastline along a place called Inaccessible Bay, due east of St. Vincent's Pass. I decided to make for this place in the hope of wangling a real bed, instead of the dubious pleasures of the bottom of the boat.

The sun was setting when we finally steered into Inaccessible Bay. It was a wide expanse of water lined by low hills; at its far end I could just make out several Quonset huts. My thoughts of a hot shower, warm food, and bed were shattered by a cry of warning from Lew. I looked ahead, expecting a log or distant reef, but saw nothing. Then I looked over the side and realized our folly. We were in only a few feet of water, and beneath our keel unbroken coral growth thrust upward. In a flash I understood how this bay had received its name.

It took us another hour to reach the end of the bay, for we had to climb out of the boat several times and drag it around large heads of coral. We finally arrived at the sandy shore in complete blackness. "Does this guy know we're coming?" Lew asked. For all I knew, he didn't even remember me. Here we were, unannounced, at the aquaculture farm of a man I had met briefly a year ago. I put on my best face and pounded on the door. After a pause the door opened and my French friend stared at us in surprise. But he soon recognized me and gave us the hearty welcome of a man who has no neighbors, a man overworked, a man at the end of his hope. He offered us omelets (not oysters) for dinner, and during a happy meal I finally asked him how his fortunes went with the oysters. His smile faded. "You will see in the morning" was all he would say.

We awoke at dawn to the sound of howling winds. Inaccessible Bay was a wilderness of waves, and our small boat thrashed at its anchor cable like an angry dog pulling at its chain. I knew we wouldn't be able to leave for several days, at least. Resigned, I began to think about oysters.

We spent the morning looking at the oyster-growing facilities. Maurice had spent considerable time and money in establishing his operation. He had imported oyster spat of a commercially grown Japanese species, already settled on empty shells, and had placed the shells in the shallow waters of the bay. During his first season his hopes soared. In the tropical waters of New Caledonia, the Japanese oyster, a species known for its delectable taste, thrived and grew prodigiously. In the cooler waters off Japan, these oysters reached full size in about three to four years; in New Caledonia they grew so quickly that Maurice knew they could be harvested during the second year. He had franc signs dancing in his eyes. But near the end of the first growing season disaster struck. With the austral summer came long heat waves. The tropical sun blasted the shallow waters of Inaccessible Bay relentlessly, and finally raised the water temperature so high that the oysters began to die. Each morning as Maurice visited his holding pens, he died a little too. None of the oysters survived into the second year.

In his second attempt Maurice tried a different tack. Oysters are usually raised in shallow waters that are commingled with freshwater runoff. But Maurice had to get his oysters into deeper water, where the temperatures would be somewhat buffered. He moved the large metal racks holding the growing oysters into subtidal depths in the bay. But in this environment a new danger lurked: predators.

I donned a face mask and fins to inspect the oysters. Their holding racks were now in depths of about ten feet. As I dived down I marveled at the amount of work that must have gone into the assembly of the racks, for each was now screened with fine wire mesh. But the screen was not enough. There was always a hole somewhere, a tear in the fabric. And even a small hole was big enough to let in monsters. I dived down again and saw the predators. They seemed innocuous enough. Under the pens I saw a jumble of scurrying crabs, nipping at the wire with their claws, searching for an entrance. Sea stars of glorious hues lounged about the screen, their seeming inactivity masking their deadly intent. Many carnivorous snails clustered around the wire as well, and as I peered through the wire mesh I could see hundreds more inside, patiently drilling oyster shells with their specialized mouth structure, called a radula. A snail would work for hours and even days until it had finally bored a tiny hole through the thick, calcareous oyster shell. With an entry way through the shell finally achieved, the voracious snail could feast on the flesh inside.

I rose to the surface, and as I took a welcome breath I was startled by a huge shape moving beneath me. A large ray had moved onto the screens, its giant batlike wings flapping gently. I could hear a scraping sound as its monstrous teeth tore at the screens. In the warm salt waters these screens, already rusting, wouldn't last long. I had no doubt that the giant rays, adapted to eat species similar to these oysters, would keep returning until they too got to the feast inside. I took a final glance at the scene below and returned to shore to commiserate with Maurice. No amount of ingenuity could save this venture. Oysters could no longer live in the sea. Their time as creatures of the shallow ocean bottom was long gone.

The Rise of the Bivalves

Of the major groups of mollusks that inhabit the world today, the bivalves were the last to appear. Although some tiny, extremely rare species are known from Cambrian strata, the group did not become common until Ordovician times, and even then they were far less common than the brachiopods. By the time of the Lower Paleozoic bivalve radiations, the brachiopods were well ensconced in almost every marine habitat. The bivalves had to fit in around the edges.

The bivalves and brachiopods derived their food in almost exactly the same way, by straining the water of its life-giving plankton. To explain the dearth of brachiopods living today in comparison with the abundance of bivalves in the world's oceans, many zoologists suggest that bivalves are more efficient at filter feeding, and so took over the place once occupied by the brachiopods: in the competition for food, the bivalves won. The fossil record, though, does not support this interpretation. Early in their evolutionary history, the bivalves were consigned to the low-rent districts of the seas, while brachiopods reveled in the richest habitats: the shallow shelf regions of the oceans. For most of the Paleozoic we find the bivalves in environments that are quite difficult to live in: shallow lagoons, where extremes of temperature can lead to quick death; in estuaries, where salinity varies; in shifting sands, where sudden storms can either unearth or bury the creatures. Most of these conditions are lethal to brachiopods. Brachiopods have such inefficient kidney systems that they cannot tolerate great swings in salinity; all but the inarticulate brachiopod genus *Lingula* (chapter 2) require water of normal salinity. Brachiopods also have the distinct disadvantage of being unable to reattach themselves if they are ripped from their anchorage. And unlike many bivalves, which can burrow with a muscular foot, brachiopods will die if they become buried in sediment, or are ripped from their attachment sites by water action. For these reasons they avoided the marginal marine sites, places readily colonized by the bivalves. It is in these environments of Paleozoic age that we find the first fossils of creatures very familiar to us: mussels, scallops, and oysters, the still-living flat clams. Their fossil shells are virtually identical to those of our present oceans. They waited out the Paleozoic Era, and when the Permian extinctions had finally swept away most of the world's brachiopods and laid open for conquest the bottoms of the shallow seas, the bivalves were poised to exploit them: they radiated into the brachiopods' old habitats.

The Permian extinctions, which closed out the Paleozoic Era, left a much-emptied world. The virtually endless expanses of shallow-shelf seas, practically deserted in earliest Triassic times, provided the opportunity for a new group of creatures. The seas became filled with plankton again, and a new assemblage of marine creatures rapidly evolved to exploit it. An entirely new assemblage of ecosystems was soon in place, dominated not by brachiopods and crinoids and archaic corals, as in the Paleozoic, but by scleractinian corals, those that still exist in our world, and by a host of newly evolved oyster and oysterlike bivalves. The Mesozoic marine world was born.

The World of the Flat Clams

Even the pioneering geologists of the early nineteenth century recognized that three very distinct marine faunas were recognizable in the various sedimentary exposures scattered about Europe. In older strata they could recognize assemblages of brachiopods and crinoids, and in newer strata they found clams and snails very similar to those living in the modern seas. And in between they found an assemblage dominated by flat, oysterlike clams. They named these huge groupings of strata the Paleozoic, Mesozoic, and Cenozoic, for old, middle, and new life. The Mesozoic was very much a time of the oysters.

Hundreds of species of oysterlike creatures evolved. They sat on the surface of the seabed, straining the rich plankton through their gills, which thus served to extract food as well as oxygen from the water. The seas become filled with these creatures, and some species became very large indeed; the shells of some oysterlike clams of the genus *Inoceramus* were over six feet long. This was the bivalves' heyday. But the golden age ended quickly. The very success of the flat clams led to their downfall. With so many of them thriving in virtually all marine habitats, the flat clams became a resource to be exploited. During the Triassic and part of the Jurassic periods, the thick shells of these flat clams protected them from the predators of their day. But the succulent, abundant flesh of these giant clams was too tempting to

Major ecological types of bivalved mollusks. The flat clams, forms that live on top of sediment or rock, are represented by Mytilus. *All others shown here are burrowing clams—forms that did not evolve until the mid-Mesozoic.*

withstand the clever advances of the evolutionary process. The key to eating these clams lay in finding some way to break through their shells and thwart the cement defenses that had proved so successful. A new type of predator was needed, a predator capable of breaking into the shells of the clams and other creatures with calcareous exoskeletons.

The Role of Predators

During the 1950s and 1960s great advances were made in the various fields known as ecology. Such concepts as diversity, competition, and predation were subjected to actual field experiments and for the first time were defined mathematically. One of the first scientists to manipulate actual habitats in order to test the validity of competing hypotheses was Robert Paine of the

University of Washington. Paine was fascinated by seashore communities and was especially intrigued by the phenomenon of intertidal zonation. Even the casual observer can see that organisms living on rocky shores exposed to the tides congregate in broad horizontal bands, assemblages of organisms stacked one upon another. On the Washington coast a broad band of barnacles overlies a distinct band of mussels, which itself overlies a more heterogeneous assemblage of seaweeds and mollusks. It had long been thought that these vertical zones reflected simply the tolerance of various organisms to exposure to air when the tide was out. Bob Paine was not content with such a simple, untested generalization: he wanted to know actual causes. He was intrigued by other questions, too. Why was it, for instance, that the zone immediately beneath the mussels contained a large variety of species while the mussel zone contained a great many individuals but only a few species? To answer such questions he performed some simple but elegant experiments. He constructed some bottomless wire cages, attached them to the rocks where the organism lived, and left them there. The cages in no way interfered with the organisms' ability to get food or to reproduce. They had but one function: to keep the creatures within safe from predators.

Bob Paine repeatedly made the long drive from Seattle to Tatoosh Island, on the Olympic Peninsula of northwestern Washington State, and over time watched astounding changes take place inside the cages. Within several weeks the composition of species within the cages began to change. In the cages attached to rocks beneath the mussel zone, regions that typically contain the highest diversity of organisms, mussels began to proliferate, covering or pushing aside the other creatures living there. Eventually only mussels could be found in these cages. When the various predators that usually preyed on the creatures in this zone could no longer enter it, so many mussels survived that they overwhelmed the other creatures, until finally only one species remained in the cage: the mussels themselves. Paine arrived at a startling conclusion: predators in an ecosystem, rather than reducing the number of species by their activities, actually *increase* the diversity of species by not letting any single species gain ascendancy. His study also showed that the mussels would be living lower in the subtidal regions if they could. They don't live in the harsh intertidal regions because they want to be there. They are there because those regions provide their best refuge from predators.

Bob Paine's studies were enormously influential for a host of reasons, not least because they pointed to the important role played by predation in the nature and structure of ecosystems.

People began to think about predation in ways that had not occurred to them before.

The Mesozoic Marine Revolution

Natural processes are a lot like trends in human welfare. In war and nature an interactive evolutionary process occurs: as one side produces new types of weapons, its enemies have to develop new defensive tactics or structures. The alternative is capitulation—or extinction. Innovations in warfare tend to make older defenses forever obsolete, even defenses that have served well for a long time. No one would recommend that we persist in making propeller-driven planes for our air forces or in dressing our soldiers in red uniforms and marching them onto the battlefield in long columns. Physical laws have not changed; biplanes can still fly and soldiers can still march. But in a modern battle such tactics are no longer effective.

So it was with the flat clams. Their sedentary life at the bottom of the sea had worked well for hundreds of millions of years. The thick, calcareous shell was an unassailable defense, for virtually no predators of the Paleozoic or early Mesozoic Era had the equipment necessary to break through it. But beginning in the Jurassic Period and accelerating through the Cretaceous, a variety of predators evolved means to get at the flesh of the flat clams. And their introduction into the marine ecosystems spelled the doom of the flat clams as assuredly as the machine gun forever ended the stately marching columns of nineteenth-century wars.

The man who first brought attention to the changes in the Mesozoic predators was Gary Vermeij, a zoologist with a keen knowledge of the fossil record. Vermeij had been studying the way crabs attacked and ate mollusks. The crabs used their claws to peel open the shells. Vermeij began to think about the evolutionary history first of crabs and then of the other creatures that he called durophagous, or shell-breaking, predators. He soon amassed a large list of such creatures, and realized that most had evolved no earlier than the Mesozoic Era. He called this evolution of new predators the Marine Mesozoic Revolution. And a revolution it was, for it completely changed the nature and composition of the marine ecosystems.

The shell breakers attacked the shelled creatures on a broad front. Among the new predators were the marine snails that evolved the ability to drill holes in shells. This capability evolved among existing stocks as well as in an entirely new order, a group called the neogastropods. These animals, which include species common to our oceans—oyster drills, cone shells, triton shells—are entirely carnivorous. They began to fill the seas with their voracious species in the Cretaceous Period and are still proliferating today, 100 million years after they first appeared.

The carnivorous snails used an esoteric method of overcoming the defense of the flat clams, for their drilling mechanism depended on a blend of mechanical and biochemical ingenuity. A tonguelike ribbon covered with microscopic teeth allows a carnivorous snail to bore a circular hole in even the thickest clam or barnacle shell. The crabs and lobsters followed another path to a clam dinner: they used brute force. The crabs so common now are relative latecomers in the sea: they first appeared in significant numbers during the Cretaceous Period. Their strong, pinching claws are perfect tools for breaking into clam shells. The clawed lobsters, which also emerged in the Mesozoic, similarly use their strong claws to break mollusk shells or peel them open.

A variety of vertebrate creatures also evolved the ability to break and feed on the flat clams. Skates and rays, which evolved from the cartilaginous sharks during the Mesozoic, feed exclusively on mollusks. They traded the sharp, pointed teeth of their shark ancestors for large, rounded ones that are useless for tearing into flesh but perfectly adapted for breaking open shells. The bony fish also evolved several groups with teeth capable of breaking open mollusk shells. Most dramatic among the Mesozoic shell breakers were the placodonts, a group of large marine reptiles capable of diving down to the sea bottom and staying there long enough to eat their fill of clams. Like the skates and rays, these reptilian carnivores had evolved specialized teeth for breaking up the shells. In all, a huge array of carnivores began to exploit the resource of the flat clams, and other mollusks as well, during the middle to late Mesozoic Era. The clams responded by producing ever larger and thicker shells and more powerful muscles to hold them closed. But the contest was never in doubt. As the Mesozoic progressed, it became clear that to be a flat clam on the sea floor was to be a meal.

The plethora of newly evolved shell-breaking predators completely changed the marine ecosystems. Clams had but one choice: respond to the predators in some new defensive way or die out. Some clam lineages responded. In mid-Jurassic time a previously obscure family of clams produced two new innovations that allowed them to burrow under sediment and still inhale sea water. A long tube functioned like a snorkel to bring fresh food and oxygen to the buried clam, and changes in the hingement and foot permitted rapid burrowing. With these innovations, increasing numbers of clams escaped the new shell-breaking predators on the surface of the seabed by fleeing into the sediment. With their long snorkel-like tubes the clams could suck in large volumes of water and nutrients, and at the first sign of disturbance they could withdraw the tube and burrow deeper into the sediment.

The adaptation worked. Although the moon snails and a few other predators followed the clams down into the sediment, to stalk them deep in the sand and mud, most predators were stymied. By the Cretaceous Period thousands of species of clams had retreated into the sea bottom to escape the shell breakers. They live there still.

And what of the flat clams? To survive they too had to find habitats that would give some protection against the shell breakers. Some, such as inoceramids, migrated ever deeper in the sea, to depths where few snails or crabs lived and where the diving reptiles of the Mesozoic could not follow. But the great depths are not favorable places for creatures that depend on plankton for food, for very little plankton lives in the dark, cold abyss, and by the end of the Cretaceous Period, some 66 million years ago, these once-common species were extinct. Other species, such as the exogyrids and gryphaeids, tried to strengthen their shells through extra thickening, but they too died out by the end of the Cretaceous. And finally, others such as the oysters and most of the mussels, survivors since the Paleozoic, were forced out of the marine realm into the brackish backwaters and estuaries or the high intertidal regions, places of temperature extremes and sudden changes of salinity—the wrong side of the tracks. Those that survived did so because they learned to live where the predators could not.

In the rocks that the pioneering geologists of the nineteenth century called Cenozoic, or belonging to the time of new life, they found fossils not too dissimilar from the creatures of our modern oceans—burrowing clams and predatory gastropods, stout crabs and sea urchins, sharks' teeth and tusk shells. What they found were the survivors of the Marine Mesozoic Revolution.

Could any of the flat clams of the Paleozoic and Mesozoic make a living in the world's oceans today? The answer to that question was revealed as a by-product of a wonderful experiment conducted by Michael LaBarbara, a paleontologist from the University of Chicago. LaBarbara has spent his entire career studying the designs of creatures both living and extinct. In his effort to understand the design of the flat clams' shells, he made casts of well-preserved fossils of several long-extinct Mesozoic oysters out of synthetic material that closely resembled the calcium carbonate of the original shells: like a modern-day Dr. Frankenstein, LaBarbara brought the shells of extinct creatures back to life, in every way save one—there was no flesh inside. He deposited these shells in a variety of shallow underwater environments and left them there. When he returned some weeks later to inspect his handiwork, he found that all of the shells had been destroyed, even though they contained no meat. The break marks on the shells clearly pointed to the culprits: crabs had discovered the odd shells and made short work of them. The crabs must have been disappointed to find nothing edible after going through an exercise that no other creatures of their kind had attempted in over 60 million years. The crabs that had attacked the flat clams of the Mesozoic and finally driven them to extinction, or at least out of the sea, were very like those of today.

Orcas Island, Washington

It's a relief to get over the side of the boat, into the cool water off Orcas Island, a high rocky massif in the green waters of Puget Sound near the Canadian border. Shafts of bright May sunshine slant downward as I move through the tangled kelp near the shore, a brown forest swaying in the current. I reach the cobbled bottom twenty feet below me and strike out to the east, following the gentle gradient into deeper water. Soon I'm crossing a vast plain of fine sand, the lazy strokes of my fins swirling sediment behind me. The bottom just beneath is pocked with countless dimples, like small coins had been pressed into the sediment surface, and as I look closer I see that each of these actually consists of two tubes set against each other, their tops covered with a ring of fine tentacles. I peer closer, my face mask now only inches from one pair of tubes, and gently poke the structure. It rapidly withdraws into the sediment. I try to grab the descending tube, but the attempt is futile; further effort will only break off the end of the neck. For this underwater field is the home of thousands of clams, buried in the sand below: large species such as the horse clams and geoducks, and smaller forms such as butter clams and cockles. The number of species here seems to be endless. Their necks, the only traces of their presence, are everywhere. Predators dot the sand: large starfish slowly cross the bottom, and an occasional bulge marks the presence of a marauding moon snail moving slowly just beneath the surface of the sand, hoping to blunder into a buried clam. But most of the clams, slowly growing over tens of years, are secure in their ability to burrow out of harm's way.

I rise off the bottom, once again moving over the vast field of clams, survivors all, direct descendants of species first evolved when dinosaurs ruled the land. And then a marvelous thing happens: whether by chance or because the clams are disturbed by the shock wave of my passage, first one clam siphon on the bottom below me, then ten, and then ten times ten and more of these siphons release clouds of white milk into the sea. In a mass frenzy the clams are spawning, releasing billions of eggs and sperm cells. Chance encounters will bring some of them together, and the fertilization of enough eggs to maintain the population depends on the power of large numbers. I've been privileged to see one of nature's most extravagant displays.

I swim out of the gamete cloud into deeper water and come at last to the place I seek. I'm 100 feet deep now, over a fine muddy bottom. The necks of clams are rarer here. I've come to see one of the last of the flat clams still living in the sea. As I gently descend to the bottom I see them lying about in profusion: scallops, one of the larger of the species still existing, with shells familiar to us all. They first evolved during the middle part of the Paleozoic Era, and exist still, on this and many other sea bottoms around the world. Nearby I see a starfish. It will have no luck with these flat clams. Unlike the oysters, now confined to the margins of the sea, or the inoceramids and exogyrids, now long dead, the scallops still thrive at the bottoms of many seas, greedily filtering the water of the most favored of marine habitats. I know the secret of their success. I rear back and kick hard with my fins, sending a shock wave over the motionless clams below me. With a rush the clams start madly clapping their valves together, jerk themselves off the bottom, and clatter away like so many sets of novelty-store false teeth. As the scallops swim to safety I head back toward my own world. Alone among the flat clams, the scallops learned to swim away in the face of danger. And so they have survived, not on the fringes of the sea but downtown, on Broadway.

4

The Kraken Wakes
Nautilus and the Rise of the Ammonites

The First of the Last Ammonites

In March the Dorset coast of southern England is a cold, stormy place. The gray sea and sky are separated only by the white froth of marching wavetops. The region's small seaside towns, standing forth bravely against the rain and crashing surf, only seem to emphasize the bleakness of this latitude. Spring is still far away when, in 1985, I step from the train into an arctic wind. I try to zip my coat yet higher and, shouldering my knapsack, strike out across the beach, toward black seaside cliffs. I am excited and chagrined. The place I approach is one of the most famous fossil localities on earth, the seaside cliffs of Lyme Regis. Here in the nineteenth century a local woman, Mary Anning, made spectacular discoveries of prehistoric reptile skeletons and fish bones—fossils that helped convince a skeptical world of the existence of the incredible, now-extinct beasts that had once inhabited this place. Mary Anning had discovered the fauna of the Mesozoic Era, the Age of Dinosaurs. John Fowles, an amateur geologist who immortalized the beach and its cliffs in his novel *The French Lieutenant's Woman,* serves as the unofficial curator of fossils at the local museum. Most important to me, one can find more ammonite fossils here than anywhere else in England, perhaps in the world. The locality I approach contains untold numbers of these spiral fossils, long ago thought to be the remains of the kraken, ancient sea monsters. These once-dominant carnivores of the sea are now completely extinct. The cliffs, the talus slopes, even the beach are littered with the fossils of these creatures; but I am woefully unprepared. I hadn't intended to visit this beach and spend a day with these old friends, but I was drawn here. I have no hammer, no chisel, none of the tools necessary to my trade. I feel naked walking out on this beach toward paleontological treasures with nothing to collect them with.

As I cross the last stretch of beach before reaching the black, rain-slick shales of the cliffs, I begin to see the fossil remains of ammonites on virtually every rock. The strata here were deposited soon after the start of the Jurassic Period, which commenced about 210 to 215 million years ago. I try to count the number of ammonite species visible in the rocks, and soon count more than

Jurassic-aged shales along the beach at Lyme Regis, on the English Channel.

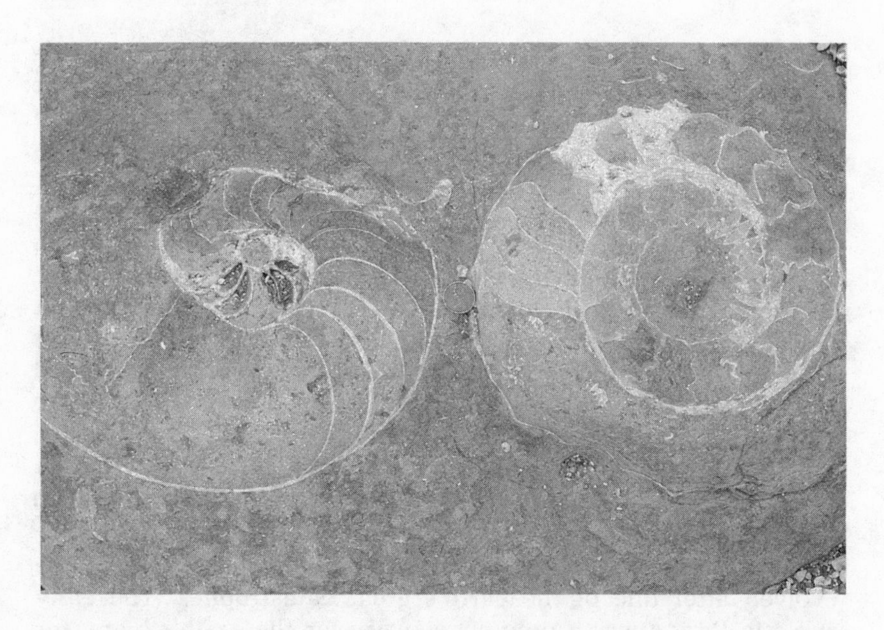

Fossils of a nautiloid (left) *and an ammonite* (right) *from the beach at Lyme Regis. Both of these individuals lived together almost 200 million years ago, and after death became entombed in the same rock. Both shells have been partially eroded away, and show the differences in complexity between the simpler nautiloid design, with its gently curved septa, and the more complex ammonite design. Both are more than a foot across in size.*

a dozen. For me the shapes and forms of these fossils are a wonder. My professional career has involved the study of ammonites far younger than these, species that lived near the end of the reign of ammonoids, species that made up the last ammonite faunas of the world's oceans before the catastrophe that snuffed out so many of the earth's creatures 66 million years ago. Most of the last ammonoids were either streamlined swimmers or passive floaters. Natural selection had favored two very different paths in the course of the ammonites' evolution, and both stocks were superbly adapted for their respective modes of life. But here on this beach, enclosed in rocks well over 100 million years older than any I have ever collected, is an almost ludicrous assemblage of species. Most look like ungainly wagon wheels. These shells, with their many-spiraled whorls, would have been most cumbersome and inefficient to move about in. Like the shell of the still-living chambered nautilus, ammonite shells contained air and water-filled partitions that gave the animal neutral buoyancy—the ability to hang weightless in the sea. Each creature was a bit like a miniature submarine, encased in its carbonate shell, able to swim or hover above the bottom as it searched for prey. For this type of life, streamlining of the shell was an important adaptation. But the shells around me are anything but streamlined. It's almost as if someone had invented a new game and decided that when play begins, any shell shape is good enough. As long as you have the basics—a shell that floats—you'll survive and flourish. But as the game goes on, you'd better improve. The assemblage of ammonites on this English beach are from early in the game and remind me, in a way, of the cars in a 1930s movie. They're charming and interesting but archaic. They'd be fun to drive along a country lane on a beautiful Sunday, but not exactly the sort of thing you'd want to have in the daily survival test on today's freeway in rush hour.

The ammonites of this beach are among the very first to have evolved after one of the earth's great catastrophes. The early Jurassic Period was a time of recovery for the earth's biota, for the immediately preceding interval of geologic time, the Triassic Period, ended in one of the great mass extinctions known in the geological record. Because this particular extinction occurred so long ago (about 215 million years), we know very little about it. We do know, however, that many thousands of species disappeared from the Earth rather rapidly. There is currently much debate as to whether the marine species and the land animals were extinguished at the same time. But it is clear that many early land vertebrates, including some dinosaurs, and numerous sea creatures, including hundreds of species of ammonites, disappeared over an interval of time no longer (and perhaps very much shorter) than a million years.

Several slapstick hours pass on the rain-swept outcrop as I try to break ammonites out from their stony cover, using rocks I've picked up on the beach for tools. But Stone Age geology soon loses its power to amuse as I destroy beautiful fossils with my bashing. As twilight gathers I finally decide to retreat from the cold beach. Striding briskly toward the cheerful glow of a nearby pub, I chance to glance downward and see an old friend on a large boulder—a huge nautiloid fossil. So much like an ammonite, this descendant of the ancestral stock of all cephalopod mollusks looks virtually identical to the living shells I will soon be catching in the Indopacific: the nautilus, last living cephalopod with an external shell. Like the ammonites I've been seeing all day, the nautiloid I stoop to examine has a spiral shell and chambers to yield buoyancy. But unlike the ammonites, it has living descendants. Two groups so alike in morphology and probably in ecology, yet one stock lives and one has died. This is a mystery worth exploring.

Robert Hooke on Extinction

I am not the first to puzzle over the very different fates of such similar creatures as nautiloids and ammonites. In the seventeenth century the great English scientist Robert Hooke pondered the nature of the spiral fossils from Dorset. His contemporaries considered fossils, including the beautiful spiraled forms from the Lyme Regis region, to be sports of nature—stones that had attained perfect regularity through means similar to those that produce crystals, perhaps. The prevailing belief was that all creatures had been created by God, and that if he had spared them at the time of Noah's flood, they must still be alive today. Aside from a belief in a world-covering flood, even scientists had no concept of extinction, and they had not the slightest comprehension of the immense age of the earth. Hooke questioned at least some of the prevailing beliefs. He correctly surmised that animals could leave records of their existence on earth through a process of fossilization, and that the stony shells so abundant in the rocky outcroppings of England's south coast were fossilized remains of long-dead animals. He even surmised that some of these fossils came from species that no longer existed on this earth. Hooke was perhaps the first scientist ever to contemplate the concept of extinction.

Hooke was particularly interested in the ammonite fossils, because he saw their great similarity to the shell of the pearly nautilus. In the mid-seventeenth century Hooke had received a great treasure: the shell of a modern, living nautilus, only recently discovered in the far western Pacific Ocean and brought back to England in a trading ship. Hooke studied the shell and made the first deductions (largely correct) about the functions of various parts. Hooke was especially interested in the chambered portions of the shell and their possible role in producing neutral buoyancy. He also noted the similarity between the nautilus shell and the ammonite fossils from Lyme Regis, and surmised that an ammonite's shell might once have functioned very much as that of the living nautilus did. But Hooke was a very careful observer, and even as he noted undoubted similarities, he noted differences as well. He was certain that these two stocks, though surely related, were not identical. The ammonite fossils, in fact, differed sufficiently from even this closest living relative to be considered an entirely separate stock of animals—and a stock that was (Hooke surmised, for the world was far from completely explored) now gone from the earth. Hooke became embroiled in a heated controversy on this subject with a contemporary "naturalist," Martin Lister, who steadfastly maintained that the ammonite fossils were inorganic "petrifactions." Hooke won his argument by producing the nautilus shell. All interested parties could see that the ammonite fossils were so like it in shell design that they had to be related to this demonstrably still-living stock of mollusks, and so must once have been alive themselves. In this way Hooke made the first telling argument that some creatures once common on earth are now extinct. Ammonites were thus the first creatures on earth to be recognized by man as extinct. Hooke even had an explanation for their extinction. Like many naturalists of his day, Hooke was sorely perplexed by the presence of marine fossils, clams and snails, embedded in rocks high above sea level—some of them even high on mountainsides. Some great upheaval must have put them there. Hooke concluded that the ammonites had been the victims of this or some other catastrophe. His best guess was that a great earthquake, far stronger than any known or experienced in his day, had killed them off. But one nagging question remained: Why weren't the nautiloids killed off as well?

The Rise of the Ammonites

The history of life can be great theater. Although some groups appeared, carved out some niche, and then disappeared without flair or dramatics, others had far more exciting careers. In the latter category are the ammonites. They evolved from other shelled mollusks, the nautiloids, in the middle of the Paleozoic Era, some 400 million years ago, and existed for over 330 million years, until their final extinction at the end of the Mesozoic Era, about 65 million years ago. But their long existence on earth by no means followed the typical evolutionary course: the fortunes of ammonites waxed and waned many times. No fewer than three times they came within a few species of extinction, only to rebound from the mass extinctions of the late Devonian, Permian, and Triassic periods and fill the seas once more with a vast diversity of forms. They were indeed a group with an instinct for grand theater. And in the best tradition of a good tale, they came from very humble origins.

Imagine an underwater world populated by a vast bestiary of crawlers and burrowers, browsers and grazers, a world dominated by arthropods and creatures of their ilk, of external skeletons and jointed legs, a world where most creatures were at most a few inches long. In this world, most life lived at the boundary between the muddy or sandy bottom and its water cover. Such a world existed about 500 million years ago, and it contained no fish, no rapid swimmers, no creatures of any kind longer than about a foot. This is the underwater world of the Cambrian Period. Our knowledge of this world comes from the study of thousands of places that contain rocks of this age in all parts of the globe. One such place, in British Columbia, Canada, has yielded more information than all of the other sites combined. This site is known as the Burgess shale.

The fossil record (if there is one) preserved in any outcrop of sedimentary strata is likely to be composed of isolated assemblages of fossil specimens, irregular snapshots of past times rather than a continuous record, and unfortunately, these glimpses of the past are quite biased, for only those creatures with readily preservable body parts, such as the shells of mollusks and the bones of vertebrates, are likely to be preserved. The vast majority of creatures on earth today (and probably in the past as well) have (or had) no hard parts; any such creatures will only rarely be found as fossils. Almost always the corpse has been quickly devoured by scavengers. In rare cases, however, usually in places where scavengers are few, such as sea or lake bottoms where little oxygen is dissolved in the water and sediment, the soft parts of the dead have been replaced by various minerals that have left a smear of carbonized film on the petrifying bottom sediment. Such environments are few, but they have yielded extraordinary insights into life of the past. They include the Solnhofen limestone, the Green River shale, and the Mazon Creek area, but perhaps the most famous of all such sites is the Burgess shale.

High on a mountainside in British Columbia, the Burgess shale has yielded an incredible assemblage of creatures. This time capsule provides our only reasonable picture of the sea life of 500 million years ago, and shows us that the marine ecosystems of the Middle Cambrian world were populated by creatures very different from those we know today. There were numerous wormlike creatures and arthropods of many kinds. Two of the most common groups of animals of today's seas, the fishes and mollusks, were poorly represented at that time. The Burgess world was a world of creepers and burrowers and feeble swimmers. Predators fierce for their time indeed existed, but if they somehow were resurrected and put into our oceans, they would probably have a very hard time getting a meal. There were no fish that could swim very fast in that world. Most life was on the bottom. The midwaters of the oceans were probably very empty.

The Burgess shale fauna, for all of the insight it has given, is still only the thinnest slice of time. In a sense it gives us a glimpse of the end of a world, for two evolutionary events completely changed the Cambrian world: the evolution of the cephalopods and that of the fish.

A Smithsonian Institution expedition to the Burgess Shale around the turn of the twentieth century. (Smithsonian Institution)

Some of the more inconspicuous inhabitants of the Cambrian-aged Burgess world were small, snail-like mollusks. Perhaps no more than a half inch long, these tiny creatures, if resurrected today, would probably give little clue that they were the forefathers of the most advanced invertebrates ever to evolve, the cephalopods, represented in today's oceans by the octopus and squid. The tiny cap-shaped shells of these early mollusks probably served the same functions as most mollusk shells today: primarily as protection against predators, as skeletal support for the internal viscera, and as an attachment site for the muscles used in creeping. But in this tiny group of Cambrian mollusks another adaptation was added to the shell: for reasons still (and probably forever) unknown to us, the rear of this tiny shell is closed off with a calcareous partition that creates a tiny liquid-filled space. Some gastropods today, such as limpets and slipper shells, have the same sort of space at the rear. The crucial step taken by the earliest ancestors of the Cephalopoda was to develop a method of replacing the liquid in the sealed-off space with gas. Once that step was taken, the scene was set for the evolution of a submarinelike buoyancy apparatus. It is hypothesized that in some small group of these snail-like creatures, the tiny sheet of calcium carbonate that sealed off the rear of the shell was left with a perforation that exposed a small portion of the animal flesh. It was then a simple step for the liquid at the rear of the shell to be removed by osmosis. Once the salt content of this seawaterlike liquid was lower than that of the animal's blood, an osmotic gradient caused the liquid to flow into the body, leaving behind a gas-filled space. This is the principle that the modern-day nautiluses and cuttlefish use, and it creates a buoyancy organ.

Such a buoyancy system must have evolved some 10 or 20 million years after the time recorded by the Burgess shale. By late Cambrian times, the shelled buoyancy organ developed by the first nautiloids had been exploited by hundreds of new species, all capable of hovering weightlessly above the bottom and perhaps swooping down on unsuspecting prey. We have no records of the soft parts of these early nautiloids, but in all probability these parts had features that would be recognizable to an anatomist familiar with today's nautiluses: tentacles, eyes, a funnel used in water-jet propulsion, and a set of large, heavily calcified jaws that look like a parrot's beak. By the end of the Cambrian Period and into the early Ordovician Period, the nautiloids had evolved a huge spectrum of shell shapes, some straight, some coiled, some like snail shells. Some were enormous: at least one was 10 feet long, and shell fragments from Sweden indicate that some monsters had shells as long as 30 feet. Terror had been unleashed in the sea. The evolution of these mobile predators may well have had some part in the shattering of the relative placidity of the Burgess shale world.

The development of the gas- and liquid-filled chamber in the shell liberated the nautiloids from the sea bottom and set in motion an evolutionary history that is still unfolding today. But the development of this buoyancy organ, so influential in the early success of the nautiloids, also imposed severe limitations. First, because the chambered shell portions contained gas at very low pressure, each shell had an implosion depth—a depth at which it would suddenly be crushed by the greater pressure of the sea. Since sea pressure increases with depth, the chambered cephalopods were limited in the depths to which they could go. A modern nautilus can descend to a depth of about 2000 feet. Any deeper and its shell implodes, instantly killing it. Since the nautilus seems to have one of the strongest shells ever evolved by a cephalopod, it appears that most chambered cephalopods lived in shallower-waters, probably at 1000 feet or less.

A second limitation of the shell system comes from the rate of shell growth. Most mollusks are capable of growing shell relatively quickly. Oysters, for example, can reach their full adult size in as little as one to two years. But more than simple growth goes into the formation of a chambered shell. After the shell has been produced, the chamber must be filled with gas. This is the process that is so time-consuming. Until the 1960s it was assumed that all chambered cephalopods, past and present, produced new chambers with gas already in place—the soft parts of the animal simply moved forward in the shell, leaving behind a gas-filled space, and this newly created space was then sealed off with a partition of calcium carbonate, called the septum. It was not until researchers actually conducted research on the still-living nautilus that the true story unfolded. A nautilus produces a new chamber at the back of its shell by moving forward, but instead of leaving behind a gas-filled compartment, it fills the space with liquid. This liquid-filled compartment is then sealed off with a calcareous septum. The liquid is evacuated by a very ingenious osmotic pump, which transfers it to the soft parts of the nautilus; from there it is excreted. This system provides a very simple and elegant solution to the buoyancy problem. But the removal of liquid from the chamber by osmosis takes time. One of the great surprises of the studies on the modern nautilus was just how much time this system takes. Even at surface depths and pressures, the removal of liquid from a full-sized chamber in a large

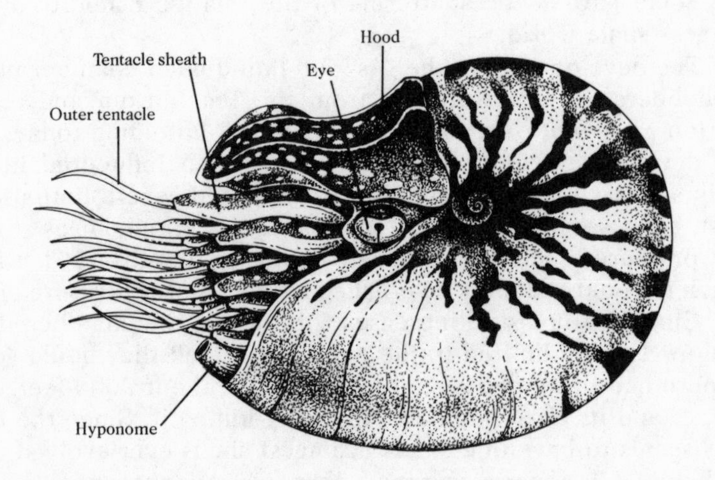

The living nautilus.

nautilus takes at least a month. And at the 900 to 1400 feet where a nautilus generally lives, the removal of liquid from a single chamber can take six months or longer. In today's oceans, an average-sized nautilus of about nine inches has between thirty and thirty-five chambers. It takes about twenty years for the nautilus to reach this size. This is a very slow growth rate for a sea creature.

Slow growth is not limited to the juvenile nautilus; it is characteristic of the developing embryo as well. Female nautiluses produce a dozen or so eggs each year, each about an inch long—very large for a creature of this size. One of the great goals of the science of embryology was to discover the way the nautilus embryos develop within these large eggs, and zoologists made many expeditions to the Pacific in the hope of obtaining fertile nautilus eggs to study. Only in recent years have such eggs been cultivated. In 1989 a team of Japanese scientists finally succeeded in hatching a nautilus, which burst from its egg looking like an inch-diameter version of an adult, with seven chambers already completed. A year had elapsed since the egg had been laid.

Several morphological changes become imprinted in the shell at the time the nautilus is hatched. These markings can be observed in the shells of fossil nautiloids as well as in those of living nautiluses, so that we can determine the size of the shells of long-extinct nautiloid species when they were hatched. The reproductive strategy of producing a few large eggs that undergo a long prehatching development is a feature shared by all nautiloids, even back to the very oldest known. Throughout their history, the nautiloids appear to have maintained a very conservative reproductive system, investing large amounts of effort and energy in a few embryos.

Three factors have thus greatly influenced the history of the nautiloids: the nature of their buoyancy system has limited the depths of the sea habitats where they can live; their growth rates are very slow, and decrease with the depth of the habitat; and each female produces very few young, which develop in their eggs for many months before they hatch.

Even with these limitations, the earliest nautiloids must have caused tremendous problems for their favored prey, the crustaceans and trilobites of the early Paleozoic seas, for soon after the nautiloids first appeared, the trilobites increased their body armor and improved their sensory equipment, especially their eyes—typical evolutionary responses to predation. Because no other creature of their world was so large and probably none was so voracious, it seems reasonable to credit the nautiloids for these developments in the trilobites. So even the slow-growing, relatively cumbersome nautiloids flourished in the Ordovician and

Silurian seas in the absence of any sort of competition. During the Devonian Period, however, the first viable threat to the nautiloids emerged. These once-invincible hunters may actually have become the hunted, especially during the long juvenile period, when they were very small. The relative tranquility of the mid-Cambrian world was shattered by the emergence not only of the cephalopods but of the other great group of marine predators, the fishes. And it is probably this group, our ancestors, that ultimately forced the ammonites to evolve.

The tiny fossils of ancestral vertebrates found in the Burgess shale look very little like the fish of today's oceans. They taper to the rear and have no bones, and the only evidence that they are indeed the ancestors of us all comes from their bilateral symmetry and the rodlike structure that runs down their backs. By latest Cambrian time these tiny protovertebrates had developed bone, but the earliest fishes did not develop a growth spurt and spring to carnivorous prominence as quickly as the nautiloids did. It appears that much of the early evolution of the fishes took place in freshwater lakes, ponds, and streams during the Ordovician and Silurian periods. While the nautiloids dominated the ecosystems of the marine realm during these early Paleozoic periods, the fish remained small, and for most of this time they didn't even have jaws; most were filter feeders, consuming particulate material or algal scum in fetid swamps. By the Devonian Period, however, the fishes began to grow larger and more numerous, largely because they evolved true jaws. They radiated explosively in the Devonian and began to fill the seas as well as the lakes; from tiny minnows giant species sprang. And it is more than likely that these new streamlined hunters found the flesh of the more slowly moving nautiloids tasty. The adult nautiloids were protected by their thick shells, but it was doubtless no great trick to break the young from their shell cover. The very slow growth of the nautiloids undoubtedly became a major liability. In any event, as the Devonian Period fish proliferated, the nautiloids dwindled.

Today the mollusks (the cephalopods, snails, clams, and two other minor classes) represent the second most diverse phylum of animals. Only the arthropods have more species; our own phylum, the chordates, ranks far down the list. The mollusks did not achieve this great diversity of forms by rolling over and going extinct in the face of an evolutionary threat. The response to the Devonian challenge of the fishes led to a major evolutionary breakthrough among the nautiloids: they produced a creature capable of competing with the fishes by tackling the three great problems of the chambered cephalopod design: growth rate, shell strength, and reproductive strategy.

Perhaps the greatest weakness of the nautiloids was their slow growth rate. When they consigned their young to a very long period of vulnerability, the nautiloid cephalopods became an extinction waiting to happen. During the earliest part of their reign, in the Ordovician and Silurian periods, the lack of mobile predators with jaw structures capable of breaking open the shell probably left the nautiloids in a very favorable position. But in the Devonian Period, when the jawed fishes evolved, the situation changed. Now the nautiloids had to find a way to grow up more quickly. Somehow they had to speed up the calcification process so that the shell would form more rapidly. The problem of accelerating the removal of liquid from the chambered shell portions was even more difficult, for that process was governed by physical laws—the same laws by which a water pump operates. If both problems were to be overcome, the shell had to be redesigned.

The chambered shell of a nautilus provides buoyancy by maintaining large gas-filled spaces that provide lift. But studies conducted on nautilus shells have shown that about 80 percent of the volume of the chambers is needed to compensate for the weight of shell; only about 20 percent provides uplift for the tissue and soft parts. The calcium carbonate shell thus accounts for most of the organism's weight. If it could somehow be made smaller, far less space would have to be devoted to uplift and far less time would be needed for growth. But the shell still has to be strong enough to withstand attack by predators and to avoid implosion under the pressure of the sea. The solution is somehow to maintain the shell's strength while reducing the calcium carbonate in it. One group of nautiloids found a way: they introduced a series of corrugations in the shell wall and especially in the septa, the partitions within the shell which intersect with the shell wall and delineate the chambers.

The septa of this new type of chambered cephalopod, which evolved in the Devonian Period, intersected the shell wall not as a straight line (as in the old nautiloids) but as a series of curves, like the arches of a suspension bridge. These curves took on the same shapes that engineers use today to reinforce structures against compression. These so-called catenary curves are optimally designed to withstand the pressure encountered by a shelled cephalopod in the sea. A thin shell and septa with such curvatures can be as strong as thicker ones and require much less calcium carbonate in their construction. And when the shell is thinner, the animal's growth rate can increase. It was this breakthrough that allowed a small group of Devonian nautiloids to avoid the fate of their ancestors. These were the first ammonites.

The internal shells of a nautilus (top) *and an ammonite* (bottom). *Shells of the ammonites differ from those of the nautiloids in two respects. The septa (chamber walls) of the nautiloid shells are smooth, curved surfaces. The septal sutures (lines along which the septa intersect the inner wall of the shell) are gentle curves. The septa of ammonite shells, on the other hand, are fluted at the periphery, and the septal sutures are folded into complex, frilled curves. The complexity of the septa and septal sutures of the ammonite shell made it possible for the shell wall to be relatively thin. The other difference between the nautiloids and the ammonites is the position of the siphuncle, the organ that empties the walled-off chambers of fluid. In the nautiloids the siphuncle generally passes through the center of the whorl; in the ammonites it is on the outer wall of the whorl.*

Paleozoic Success and Failure

The reduction of shell by the ammonites was the start of a trend that has culminated in the cephalopods so familiar to us today. Octopus and squid have no external shells of any sort, and it seems odd to us to realize that the shelled nautilus is related to these fast-moving, streamlined creatures. Yet when we examine the entire history of the cephalopods, we can see that the aberrations are the nautiluses we know today, not the octopus and squid.

The new design of their shells allowed the ammonites to grow faster than nautiloids with shells of equal diameter. And with their much faster growth rates, the ammonites were vulnerable to predation for a much shorter time. But the final evolutionary step that produced ammonites out of nautiloids had little to do with growth rates: these creatures also solved the challenge of their slow prehatching development and low fecundity. The ammonites evolved an entirely different reproductive strategy, producing vast numbers of very tiny eggs.

Producing many eggs and broadcasting them into the water, or having them hatch into tiny larvae that then spend some period of time in the plankton, makes very good evolutionary sense in some respects. During times of environmental stress, it is often advantageous to have a mechanism that allows the young to escape in the plankton. Oysters and many other creatures are incapable of moving about, but they can disperse to widely separated parts of the globe because their eggs and newly hatched larvae, floating for days or weeks in the plankton, are carried by the ocean currents for vast distances. Even if some change in the local environment—the introduction of toxins, say, or a rise in temperature—dooms the parents, the larvae may be able to float away from trouble, eventually settle elsewhere in a more favorable setting, and keep the race alive. This seems to have been the case with the ammonites. Where they hatched, their shells were about one millimeter in diameter. Twenty-five of these tiny hatchlings stretched together, shell to shell, would be needed to cover an inch—the length of a single nautilus hatchling. Although we have no direct evidence, it is probable that as the average size of an ammonite egg decreased, the number of eggs a female produced increased. Instead of the dozen or so eggs produced every year by the nautilus (and probably by the extinct nautiloid species as well), each female ammonite probably produced thousands or tens of thousands of eggs.

These two fundamental changes in ammonite morphology—the reduction of shell material through improved design and the reorganization of reproductive strategy—set the ammonites completely apart from their nautiloid ancestors. The effects were immediate. While the number of nautiloids dwindled, the ammonites radiated explosively into many hundreds of new species. Even in the tempo of their evolution the ammonites differed markedly from the nautiloids. Whereas nautiloid species lasted a relatively long time, the ammonites evolved and were extinguished very quickly. Many new species were constantly appearing, only to die out after a short time, for reasons still unknown to us. This susceptibility to extinction has made the ammonite species useful timekeepers, for their fossils enable us to know the ages of rocks in which they are found.

During the 20 million years that followed their first evolution, the ammonite tribe evolved into 25 genera and hundreds of species. These early ammonites, sharing the oceans with the nautiloids, still made up only about 20 percent of all cephalopods. Then the ammonites were devastated by three crises about 7 million years apart. Each of these crises reduced the current ammonite population by more than 90 percent. After each pulse of extinction the surviving ammonites would speciate anew, rapidly producing many new taxa, only to be knocked down by the next extinction episode. We don't know with any certainty what caused these extinctions. Some scientists suspect they were caused by a rapid change in either sea level or climate; others believe that a giant meteor hit the earth. Whatever the reason, the ammonites weren't the only species affected; many species of corals, brachiopods, and fish went extinct as well. The last of these crises, which came at the end of the Devonian Period, was the most severe for the ammonites: of eighty genera, only two survived. But those two remaining ammonite taxa made the most of an empty world. They rapidly evolved into numerous taxa, and this time they were rewarded with a long period of relative stability. For more than 100 million years the ancestors of those two survivors of the lost Devonian extinction evolved with breathtaking rapidity into several hundred genera and thousands of species. Meanwhile, the nautiloids had continued to wither, until very few taxa were left among the thousands of ammonite species in the seas.

The long dominance of the ammonites seemed to be ended forever near the close of the Permian Period, about 225 million years ago. During the last few million years of the Paleozoic Era, the greatest single mass extinction recorded in the fossil record unfolded. For reasons that are still unclear, but that are thought to relate to changes in sea level and climate associated with the convergence of all the continents into a single huge land mass, extinctions of unprecedented magnitude took place in the sea and on land. It has been estimated that as many as 90 percent of all species—*all* species—became extinct at this time. The earth has known no greater disaster than this one. We break up geological time into major units, largely on the basis of fossils. The mass extinction at the end of the Permian Period was so destructive that it signals the end of an era—the Paleozoic Era.

It is tempting to speculate about Day 1 of the Mesozoic Era—the first day of the Triassic Period, the day after the end of the extinctions that closed out the Paleozoic Era. But such a concept is meaningless, for no scientist believes that the final Paleozoic extinction happened suddenly. There are no scenarios involving a spectacular meteor hit on the earth, for instance. It is believed that this wholesale destruction of the earth's biota took at least several million years to complete. Over this long period species after species gradually shrank until each was represented by just one population, and a day came when the last member of that population died. When the episode was over, perhaps only 50,000 of an original fauna and flora of well over a million species were left on earth. There was no Day 1 of the Triassic that we can define, even in the abstract. There was only a long slide of ever-declining numbers of species until some minimum number was reached. The Triassic Period, which is the start of the Mesozoic Era, began when that long decline finally bottomed out and the number of species finally began to increase again.

Though the variety of creatures in the world was relatively small at the end, the *number* of creatures may have been substantial. The few species that survived the crisis may have maintained huge populations. The whole world may have been like today's Arctic regions, where we may see large numbers of individual creatures but a very small number of species. But a dull world it would have been, at least to my mind. There were no coral reefs, for instance. There was little diversity of fauna on land. The sandy and muddy bottoms of the sea would have seemed quite impoverished in comparison with our teeming world. And swimming about in that lonesome sea were only a few species of chambered cephalopods—perhaps one or two species of ammonites and a couple of nautiloids, all that remained of the wide variety of these two stocks that existed at the end of the Paleozoic Era.

Mesozoic Rebound

The great trick in the history of life apparently lies in being able to survive a mass extinction. If you make it through, you inherit a world empty of competitors. And just as the first nautiloids were able to speciate rapidly in the absence of competition at the end of the Cambrian Period, the ammonite survivors of the earliest Triassic Period soon filled the seas with a profusion of shapes and forms, far more species than their Paleozoic ancestors had produced. And the seeds of this great outpouring of species were those one or two survivors of the great Permian extinction.

The Triassic Period records one of the shortest distinct groupings of life known in our rock record. Despite the several mass extinctions of the Paleozoic world, its faunal assemblage as a whole is recognizable for several hundreds of millions of years. The Triassic world experienced mass extinction, but this latter extinction brought the period to a quick end. Unlike the final Paleozoic episode, it appears to have happened suddenly. There is evidence that it may have been triggered by a catastrophe in eastern Canada, for a very suspicious hole in the earth there is of the same age as strata deposited at the end of the Triassic. This huge impact crater, the Manicouagan, is thought to have been created when a huge meteor hit the earth some 215 million years ago.

The Triassic faunas, both on land and in the sea, would have seemed very strange to us. The dinosaurs were not yet established as a dominant group; there were a few of them, but far fewer than of the mammal-like reptiles—large, lumbering creatures that were the immediate ancestors of the mammals. Most of these creatures disappeared in the extinctions that brought the Triassic Period to a close. Perhaps if there had been no Triassic extinction, the dinosaurs would never have become dominant, for they flourished only after the mammal-like reptiles had disappeared.

The Triassic faunas were equally archaic in the seas. Perhaps the only familiar biota was the coral reef community, for the Triassic was the time of the first proliferation of scleractinian corals, the same group that today make up the coral reefs of the tropics. Other communities would have seemed more bizarre. With the disappearance of the brachiopods at the end of the Permian, other creatures radiated to fill the void. In Triassic times the most successful of these creatures were bivalved mollusks. Although some of these clams, such as mussels and scallops, would seem somewhat familiar to us, the majority of Triassic clams would seem curious indeed. Over 50 percent of them died out at the end of the Triassic Period. The ammonites were also common in the Triassic marine realm, and they too suffered a spectacular extinction. During the Triassic time span they had radiated to well over 400 genera and thousands of species; at most one or two species survived the crisis that ended the Triassic. Once again, as at the end of the Permian, the incredible variety of ammonites dwindled to a handful of species. And once again the world was a lonely place.

An artist's version of ammonites as they may have looked in a Mesozoic sea.

The expansion of ammonite species in the early part of the Jurassic Period, some 180 million years ago, continued the pattern that had followed earlier near-extinctions: new species were produced so rapidly that they came to outnumber the Triassic species. More ammonite species flourished during the Jurassic Period and the following Cretaceous Period than at any other time during the 300-million-year reign of these creatures. In size they ranged from dwarfs perhaps half an inch long to giants with shells more than ten feet in diameter. The shell shapes evolved by the Jurassic and Cretaceous ammonite stocks showed a bewildering variety of forms and patterns. The familiar flattened spiral patterns of all previous ammonite species were augmented by a profuse variety of uncoiled shells, ranging from straight shafts to snail-like shapes. The ammonites shared the seas with archaic fishes and great reptiles, while the dinosaurs ruled the land.

By the early Cretaceous Period, about 125 million years ago, the ammonites were among the most common creatures of the sea. They were particularly abundant on the shallow continental shelves, on the warm sandy or muddy bottoms where life has always been most abundant. But sweeping changes were under way in the marine ecosystems: new types of predators were evolving to exploit the long-unassailable ammonites. For several hundred million years the ammonites' shells had protected them from the ravages of the many other carnivorous creatures of the sea. In the early Cretaceous the key to the ammonites' citadel was finally found. Like the flat clams that lived among them, the ammonites became victims of the Marine Mesozoic Revolution.

Palau, Micronesia

The clear water of Palau has a subtle, crystalline look that I have seen nowhere else. The sea is an aquamarine; you dive into an otherworldly jewel.

I pass downward through this splendor quickly, for once ignoring the fragile reef wall beside me, just watching my depth gauge and keeping the nautilus I hold in each hand oriented so that they can breathe. I need to get the two animals down to cooler water as quickly as possible, for the blood-warm sea around me can prove fatal to them after even a few minutes of exposure. My buddy Mike Weekly is at my side. I can't see him but I feel him and hear him. We fall through the 100-foot mark, still descending. This part of the outer reef drops in a sheer vertical wall to the ocean floor. The blue below me turns black in the infinite distance. I feel like an astronaut, weightless. My senses tell me that I'm in air, not water, and I should be falling toward the darkness below me, not floating gently. I should be more on edge, for downward is death for me—and just the opposite for the two creatures I carry, who are under a death sentence in the absurdly shallow waters to which I must confine myself. Laws of physics and the work of gas molecules are my masters in this world, restricting my underwater voyages. I caught three nautiluses earlier this morning. I outfitted one of them with an ultrasonic transmitter, so that over the next week I can follow it and study its movements. But the two nautiluses I now carry are excess, lured into a baited trap placed 1000 feet below the light-filled shallows, supreme survivors fooled by human subterfuge. They are not part of my plans; they are unwanted guests now being shepherded back to the depths, for I can dive downward much faster than they—perhaps my only underwater skill superior to theirs. There are other reasons to take the nautiluses deep to be released, for it's daytime in the upper reef, where the hunters use vision to find their prey. The nautilus and its kind have been obsolete in these depths for over 70 million years, but the wanted posters are still up: nature makes no distinctions once the death sentence is passed. In the shallows the shell of the nautilus is no longer adequate protection against predators.

Mike and I stop at 150 feet and I release the nautiluses. They hang for a moment, still tightly closed in their shells, and then cautiously extend tentacles outward, testing the water. No longer feeling the vibrations caused by my grip during the descent, they extend fully and expel jets of water from their funnels. In a rush they shoot downward, then descend in huge slow spirals. I followed their part with my eyes until they become small white crescents against the blackness below. I know the bottom is about 600 feet deep here, and the nautiluses won't reach it until well after Mike and I are safe in our boat again.

I look at Mike and see the message in his eyes; he knows it's time for us to go. We're in water far too deep for our kind, and the time bomb of nitrogen is dissolving in our tissues. But as I begin to rise toward the light, I hear Mike hoot. I turn to him and see him pointing downward at the still-visible nautiluses. They're now perhaps 30 or 40 feet below us, and I see what Mike has pointed at. A larger triggerfish has appeared from the reef wall and is rapidly closing in on one of the nautiluses. The fish is brightly colored and has a clown face that masks its deadly intent. My first reaction is curiosity, for I've never seen a fish approach a nautilus before. My curiosity turns to horror as I watch the triggerfish torpedo in on one of the descending pair and smash into the shell. The huge, parrotlike jaws grab the outer shell of the nautilus and rip off a large jagged chunk; the snap of the cracking shell tweaks our ears. The triggerfish quickly finishes its work, ripping into the soft parts now exposed. With the first bite the nautilus literally explodes. A cloud of blue smoke envelops the entwined pair, the blood of the nautilus giving sad testimony to the efficiency of this particular shell-breaking predator. My first reaction is to descend quickly and intercede, but such a move would be foolish—too late for the nautilus, lethal for me. I turn away, not wanting to know if the triggerfish, after devouring the first nautilus, will turn to the second. As I rise into the shallows I'm a little sick at heart, two more nautiluses now on my conscience.

Long after I'm safe ashore the episode haunts me. For most of the long reign of the nautiluses and ammonites, creatures such as triggerfish didn't exist. They're relative latecomers on the scene, creatures of the Cretaceous Period—one of the teleost fish, which make up most of the modern-day fish fauna. Some of these fish with bony skeletons, as the triggerfish just demonstrated, evolved methods to crack even the strongest shell, and their appearance in the late Cretaceous seas in no small way helped seal the fate of the ammonites.

The nautilus can no longer live in the shallows; it can make only brief visits into this energy- and nutrient-rich domain under the cover of darkness, when the visually guided predators are rendered harmless by the night. It wasn't always this way; the shallows were once the kingdom of the chambered cephalopods. But as the new predators appeared on the scene during the Cretaceous Period, the shallow-water nautiluses and ammonites had to adapt or die. At first they tried to adapt: they built sturdier shells, replete with spires and tubercles, defenses designed to withstand the jaws of predators. But even the spiniest shells were being broken. Finally, like the Maginot Line, this static defense proved inadequate. Species after species of ammonites disappeared as the Cretaceous Period waned. There was only one escape for the ones that still lived, the same strategy used by the nautilus today: they retreated to the deep, where the eternal blackness can thwart the search of the hungriest visually oriented predator. But the deep exacts its own price: slow growth, little food. It's a draconian solution, and it put the chambered cephalopods in a no-win situation. I was reminded of this a year later, on a dark cloudy day, in haunted territory, on a long hike through the Sierra foothills of California.

The Land of Ishi

When the gold miners had finished their ravaging of the streams and rivers of central and northern California in the mid-nineteenth century, they left behind a landscape fouled by mountains of gravel and heavily dredged river courses. But as gold fever subsided, wiser heads saw that a far greater treasure than all the gold in the mother lode was spread out for all to see: the rich soil and long growing season made the northern part of the Great Valley an agricultural treasure beyond measure. Vast farms soon began to dot the valley, spreading upward on the Sierra foothills. The men who claimed and tamed the flat valley must have led a hard life, as all farmers do, and surely relished the relief of weekend hunting among the abundant deer and bears of the region. But some of those farmers and pioneers pursued other game: they went hunting for Indians. Entire tribes were slaughtered. By the start of the twentieth century only one member of the tribe that had long inhabited the region immediately south of Mount Lassen was left. His name was Ishi. To survive he moved into very rugged terrain along a stretch of white water that the white settlers called Mill Creek. Finally, perhaps in despair or out of sheer loneliness, he left his wilderness preserve and ended up in Berkeley, the private experiment of anthropologists there.

I listen to this story as I hike down toward Mill Creek on a late-winter day in 1984. Jim Haggart, my student at the time, is accompanying me on a long sampling expedition. It's a six-mile hike each way, all downhill in the morning but all uphill when it matters, with backpacks filled and heavy on the way home. We have come to collect microfossil samples from the Cretaceous-aged strata that line this twisting stretch of wild water. We're in country that normally would be open to view, the Great Valley spread out before us to the west, but clouds fill the valley this day. As we climb down toward the creek we finally find the sedimentary rock we've come to sample. As elsewhere on the eastern side of the Sierras, the older Cretaceous strata underlie the thick igneous rock of the region, and can be found only in the beds of the creeks and rivers. We travel silently, at first; although this is Forest Service land, it is still agricultural country, and so far from the main roads that California's reputedly largest single cash crop is furtively grown along the hillsides, the green marijuana growing in small glades, and we are probably watched by armed men as we pass by. We have reason to be careful; the summer before we had been confronted by a shotgun-wielding guardian on a creek several miles to the south, and had been chased off yet another outcrop by Doberman Pincers.

The Cretaceous-aged strata exposed along Mill Creek were deposited at the edge of the vast sea that stretched westward from here; the sandstones and shales we traverse were laid down in water no deeper than 100 feet, where a thriving community of mollusks and other creatures lived in warm, sunlit waters between 90 and 80 million years ago. Through chance the empty shells of

Cretaceous-aged exposures along Mill Creek, California.

the creatures that once lived in this ancient sea are exquisitely preserved. They were buried soon after the creatures died, and when they're exhumed from their stony graves they look as if they came from still-living creatures, not from species extinct for many millions of years. The community that lived here in the Cretaceous sea was made up of clams and snails, sharks and other fish, and ammonites—wondrous ammonites of many species, with golden, iridescent shells fit for museums. The Mill Creek region is the site of Jim's research for his doctoral thesis, and in the four years of his study he will eventually collect many hundreds of ammonites dating back to near the end of the Cretaceous Period.

As we climb down along the creek we move up through time, for the strata were uplifted when the Sierra Nevadas were thrust upward. The sedimentary rocks on Mill Creek are now tilted toward the west, in the direction we have descended. We become time travelers, starting in the basal stratum at the trailhead in rocks about 90 million years old and ending in rocks 10 million years younger. And as we make this voyage we pass through a time period critical to the history of the ammonites, a sad time for anyone nostalgic enough to mourn the passing of these fantastic animals. The layered rocks hold the record of a last stand of ammonites in the most favorable of all marine habitats, the shallow continental shelf, where food is rich and abundant but where the shell-breaking predators lived then, and still do. In these rocks we see the sad record of the ammonites' failure. They lived like armored tanks on the bottom, encumbered by heavy shells. The light shells they had once evolved to defy gravity no longer served their new purpose; now they needed armor. We pass through this graveyard, seeing the remains of broken spiny shells. Jim stops at the sight of a beautifully colored ammonite loose in the creek bed. We look at it closely and see the familiar breaks still etched on this ancient fossil shell. Almost every ammonite at Mill Creek was broken long ago by the predators of the Cretaceous Period. The spines and thick shells couldn't save them. The slow-moving ammonites became the prey of crabs and sharks, and of diving reptiles such as mosasaurs. As we make our way into ever-younger strata, the ammonites become rarer and the species change. In the youngest stratum we find only compressed, streamlined ammonites, the remains of the swiftest swimmers. The bulky, armored species are gone.

At midday we stretch out on the smooth rocks for lunch, our legs tingling from the long hike, and I shudder at the thought of the return trip. We're in a large natural hollow in the grainy sandstones called Kingsley Cave. Legend has it that Ishi lived here for years, watching in silence as white settlers hunted two- and four-legged victims in the area and cleared the grasslands of the

A fossil nautiloid found in the Mill Creek strata.

valley for their farms. Jim and I are nearly silent over lunch. The low-lying black clouds dampen our clothes and our spirits. In this pervasive gloom I feel as though I were being watched.

The hike back up the long trail isn't as bad as I feared, but we seem to be accompanied by nagging dread; the knowledge of slaughter and senseless death is heavy in this place. By late afternoon we're nearly back to the trailhead. We deserve a break, we decide. We off-shoulder our packs, keeping only hammer and chisel. We both love the thrill of discovery, for sensational fossils are to be found here. We seek that moment of joy when a large concretion finally splits to reveal treasure inside. We're at the side of a large cliff of sedimentary strata, and fossils are everywhere. Giant flat clams make ledges in the strata, and small clams and snail species litter the rock surfaces. But we search for the ammonites, for the most wondrous and bizarre species can be found here. For most of their history all ammonite shells were flat spirals. Near the end of their long reign, however, the ammonites experimented with other shell shapes under the selective pressure brought to bear by the merciless predators. Some ammonites evolved snail-like shapes. Others built long, perfectly straight shells; they must have maintained a vertical position in the sea. Some became giants, with shells like huge wagon wheels, six feet or more in diameter, and covered with thick ribs and spines. Others took the opposite tack, becoming very small, an inch or two at most, and perhaps lived at the surface of the sea rather than on the murderous bottom. But the extravagant diversity of form is a mark of desperation, an attempt to find some way to win under the new rules. Ammonites first evolved in a world where they were unassailable; but in these Mill Creek rocks, some 300 million years younger than those Devonian-aged creatures, the biological realities of the marine world had greatly changed. Perhaps the wonder isn't that ammonites are no longer living but that they survived as long as they did.

Jim and I are settled into the side of the large outcrop, about six feet apart. We're both covered with mud and sweat and rock dust as we attack the stony concretions with our hammers. I've found what appears to be a large ammonite in the side of the cliff. Only a small piece of it was exposed, but enough to give away its presence. I reflect on the fantastic chance that any fossil will erode out of a cliff at the exact moment when a paleontologist happens to wander by, for the specimen I remove has been entombed in this rock for over 80 million years. It's slow work. Jim is similarly engrossed, and from the ringing blows it seems that he too has discovered a prize. I've nearly freed the large discoid shell from its rock tomb when a hammer blow strikes my arm. Has Jim gone mad? I look up and see him seated still, oblivious of my shock. I look back to see rich arterial blood flowing onto the outcrop from a perfectly circular hole in my arm. Jim turns to me in the sudden silence and then looks at his hammer. One of its blows on his chisel has broken off a metallic edge from the hammer, shrapnel now embedded in my arm. My ammonite is now blood red, and I feel surrounded by suddenly reincarnated ghosts, ammonite and human, both victims of vertebrate predators. I carry the scar from Mill Creek still, a red badge of curiosity.

I'm heavily bandaged as we finally drive out, bumping over the abominable road in the worst of humors. Only half joking, I've informed Jim that attempted murder of one's supervisor is no way to earn a Ph.D. We finally come out of the cloudbank that has stalked us all day and see the valley laid out before us. I try to imagine it as a giant seaway between beaches high on the sides of the present-day mountains, the ocean bottom a thousand feet deep, there where denuded trees cover the valley floor below us. And that bottom, so far beneath us, is where the last ammonites must have been, I muse, after the massacre in the shallow bottoms of the world in Mill Creek time. The ammonites were largely gone from the shallows by 70 million years ago; like the flat clams, they were victims of the modernization of the marine ecosystems. Only one act remained to be played, the final scene of the tragedy. The ammonites took their last stand not in the shallows but in the deep. There is no sedimentary record of this final scene in California. To find the final ammonites one has to travel to another beach, near a tiny Basque town called Zumaya.

5

Death of the Polypi

Nautilus and the Last Ammonites

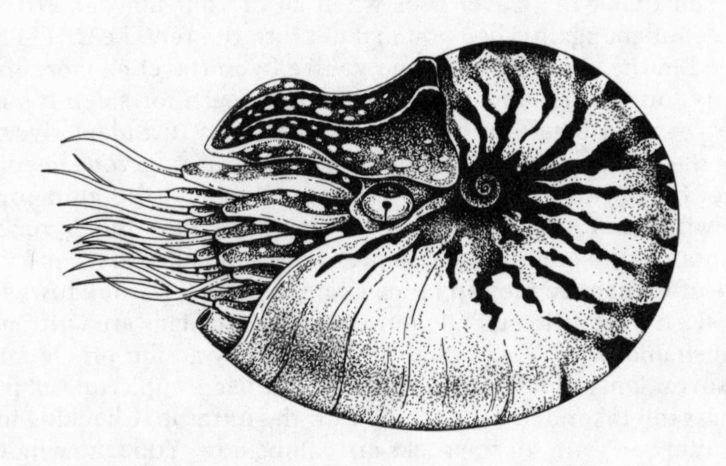

The Cliffs of Zumaya

I believe we all have places we hold sacred, places we keep in our hearts, places that retain a piece of us just as we preserve the essence of them. I keep such a place within me. It's a magical place named Zumaya. It's a place where the crashing sea meets a rocky coast beautiful beyond belief. But the world has thousands of miles of scenic coastline. I return to Zumaya because it's the grave site of the last ammonites.

Zumaya isn't an easy place to get to. You start at New York's JFK airport, probably at night, perhaps already well traveled. There you find your flight to Madrid. Aboard the plane at last, you try to sleep. You arrive with the dawn and thankfully stretch cramped legs as you queue up in one of the long customs lines. Fatigue is etched on all of the faces around you, but excitement too. You have five hours to kill in the Madrid airport, nodding off on plastic couches, unless you want to chance a dash into the city and back again in time for your connecting flight. It's early afternoon when you finally take off on the small Aviaco propeller plane, surprised that any such relics still fly, and head north and east across the flatness of Spain until, an hour later, mountains come into view, and then the sea. The plane begins a steep descent, and all the Spanish people around you light up one last cigarette, for no-smoking sections don't seem to exist in this country. And then you're swooping low over the sea, and your last glimpse before landing in San Sebastián is of a rocky coastline of steeply tilted red and brown strata, and the excitement builds again, for you've just seen the rocks you've come such a long way to study.

You thank the travel gods when all of your luggage arrives, and genuflect again when you make it into the rental car, all bags stuffed in the back. By this time you've been traveling more than twenty-four hours, and even though you yearn for sleep it's not time yet. You blast out of San Sebastián in a turbulent river of cars, the Spanish drivers careening around you, everyone heading somewhere with urgency and purpose. You find the auto route and with your tiny engine screaming head into the Pyrenees mountains, the land of the Basques. The sun is brilliant, and you drive at great speed for an hour, glimpsing small Basque towns as they flash by, until you find your exit. All road signs are written in Spanish and Basque, and most are covered by graffiti proclaiming the sovereignty of ETA, the Basque separatist group. You roll past the last toll taker and head straight for the outcrop. Checking into the hotel can wait, for the rocks are calling now. You take winding roads along a river and slow down as you approach the town you've traveled so far to see: Zumaya. An ancient fishing village, it bustles with commerce. Dark-haired women throng the market stands, and tiny, hunched men in small black berets, slow in their old age, pause to chat together. There is no neon here, no McDonald's; there is only age and dignity, and coldness too. You are noted, your license plate is seen; you are tolerated but not acknowledged. No one speaks English but French will do, and in some ways French is better to use here than Spanish, for these are Basques, not Spaniards.

You take a winding road through the middle of the town past a series of garages built into high stone walls, and the road gets ever narrower. A Basque steps from the gloom at the sound of your car, squints through you for a moment, and then returns to his industry. You emerge from the town, pass by the timeless church, and follow a dirt lane ambling between two stone walls. You bump along the rocky track, past a large field of grass worked by dour men with scythes and donkey-drawn carts, past large haystacks the shape of beehives, until you reach the edge of a cliff. And there your long voyage finally ends, before a breathtaking vista of sea and shore.

The first thing that strikes the eye is the nature of the rock. Tilted on their sides at a rakish angle are giant sheets of rock, sedimentary strata layered one upon another, shooting out to sea to be met by the crashing surf. The providential play of the tectonic forces that produced the Pyrenees has lifted these ancient strata from their subterranean resting places. They are exposed like a book thrown open, the pages beckoning, each stratum a page of time open for your reading, inviting your study.

The fantastic cliffs that make up the Zumaya exposures are among the few places on earth where sedimentary strata deposited just before the end of the Mesozoic Era, the Age of Dinosaurs, are in contact with strata that record the first moments of the Cenozoic Era, our era, the time of mammals. It's a time of great interest to students of evolution, for one of the greatest extinctions of all time occurred at this boundary. It makes up one of the most fascinating and controversial scientific topics of our time. About 66 million years ago, all dinosaurs were killed off, and many marine creatures as well, the most important being the ammonites, creatures related to the octopus and squid, the "polypi" of the ancient naturalists. Questions abound. How rapid were these extinctions? Were they the culmination of several millions of years of environmental deterioration that caused the stocks of dinosaurs and marine life to dwindle? Or did they happen much more suddenly, over thousands rather than millions of years? Or perhaps more rapidly still, over a hundred years, or ten, or one? Or over several months or weeks? Or did they happen in a day, an hour? Were the extinctions simultaneous all over the earth, or did they happen at different times in different places, starting, say, in the equatorial regions, and then spreading to the poles? Did the extinctions happen at the same time on land as in the sea? Were all types of organisms affected, or were only certain types vulnerable? Were the extinctions brought about by a multitude of factors, or by one? The answers, if they exist at all, lie in the only book of ages kept by our earth, the sedimentary record. But the critical pages of this venerable manuscript are few and must be read carefully.

Entry to the sedimentary book of Zumaya is gained by an old set of stone stairs that lead down to a tiny beach. The ocean is the true master of the Zumaya exposures, for only at low tide can you pass into the critical parts of the coastline. The time on the Zumaya outcrop is thus very limited, and woe to the intruder who fails to heed the turn of the tide. At most you have six hours to work, and usually less.

The coastline at Zumaya, in northern Spain. The tilted Cretaceous strata are exposed in magnificent fashion here. The stairway to the beach is just out of sight to the left. This picture was taken at low tide; all of the rocks beneath the cliffs are covered by the sea at high tide.

The stairway is built into the edge of a high wall of marl, a rock formed from the accumulation of mud and skeletons of microscopic plankton. This sediment was deposited at the bottom of an ocean that may have been more than 1000 feet deep. This exposed ocean bed makes up the southern wall of a giant bay. After traversing this long wall, you can pass over a perilous rocky point and drop down through time, stratum by stratum, into ever-older periods. Although the coastline in this region records more than 80 million years of earth history, only a small interval of this huge expanse of time is critical to the supplicant seeking answers to questions about extinctions. Much of the rock along the Basque coast bears no fossils and thus has no interest for the paleontologist, for the fossils are the words that tell the story; strata without them are impossible to read. At Zumaya about 200 meters of strata contain fossils. Six hundred feet of these accumulated sediments were deposited over 2 million years at the end of the Age of Dinosaurs; 150 feet more were deposited during the first million years of the Cenozoic era.

To the north the strata stretch upward, tilted at an angle of about 45 degrees. You have to clamber over sharp rocky ridges to move up through time. These are the strata that are so rarely found on this earth. Several hundred individual layers, varying in thickness from as little as several inches to about a foot, make up the wide bay and lead to another sharp rocky point. At low tide you can climb over these last rocks of the Cretaceous Period to find one last bay: Boundary Bay, the gate between the eras.

The Cretaceous strata exposed along the coast at Zumaya are richly colored, varying from tan to deep maroon. The many hues of these strata are probably related to subtle changes in the mineralogy of the sediments over the long years of their burial. As you move from the stairway, across the strata and up through time, the color stays a rich purple-brown. But in Boundary Bay, as you cross the promontory that separates this inlet from Stairway Bay, you are immediately struck by a dramatic change: the purple marls, so uniform in color and thickness, are overlain by a spectacular assemblage of more thickly bedded strata of bright pink

Cretaceous-aged strata exposed at Zumaya. The rocks become younger as one moves from right to left.

and white. The effect is stunning. With the first view of the strata in Boundary Bay you know that something dramatic happened here. So profound a change in both the color and the thickness of the rocks could not have occurred unless the environmental conditions that controlled sedimentation were themselves drastically altered.

You stand in Boundary Bay, the high cliffs towering overhead, and touch the two eras. A warm sun sends wisps of steam up from the rocks recently uncovered by the retreating tide, while the scream of gulls and lap of waves soothe your jet lag. Boundary Bay is about 50 feet wide, defined by walls of Cretaceous rock on its south side and Tertiary rock to the north. You have the last 50 feet of Cretaceous sedimentation to study, so you unshoulder your backpack and begin to search the rock for signs of ancient life.

The southern wall of the bay is made up of interbedded limestone and shale, each couplet about a half foot thick. With hammer and chisel slung from your webbed belt you scale the sides of this wall, the soles of your rock-climbing boots gripping the surface as you carefully pull your way along the wall, eyes staring at the rock, searching for the telltale spiral or curve of shell that marks a fossil. And fossils you find—mostly ammonites, some as large as a dinner plate but most less than an inch long. Here and there you see the prints of small clams, and occasionally the rounded hemisphere of a sea urchin. You collect each one with hammer and chisel, and glue back together those that break. Each fossil is a prize, to be labeled with its position, the date of its discovery, and, most important of all, the stratigraphic level at which it was found. These treasures carefully wrapped in sample bags, you continue your search. The fossils are randomly scattered through the strata and are never common, so you try to search systematically. If significant time goes by without a find, you wonder if the problem lies in the rocks or in your head, and you often find that you've been daydreaming, staring at the rocks but seeing nothing.

You search in ever-higher strata, moving upward in time, and the ammonites become very hard to find. About 30 feet below the boundary you find one more specimen, and then no more. The only sign of life now is an occasional sea urchin or clam. Finally you approach the highest exposures of the wine-red Cretaceous sediment. You can put your hand on the last grains and stare in wonder at the overlying rock. You're now looking at the boundary between the Mesozoic and Cenozoic eras.

You see a layer of clay about a foot thick. It's much softer than the underlying red Cretaceous limestone and shale, and different in color too. You break off a piece of the highest Cretaceous rocks and with your hand lens scrutinize it carefully. Under a 10X loup you can see numerous calcareous spirals in the rock, the skeletons of single-celled planktonic organisms called foraminiferans. You examine the clay layer and find it empty of these planktonic fossils; the clay layer you look at has been called the "magic layer," for it's filled with very strange material. If you have the right machines, you will find within this clay layer the metallic elements iridium and platinum in extremely high concentrations, and other evidence of catastrophe as well. It was the analysis of a very similar layer, found between the strata of Mesozoic and Cenozoic age in Italy, that led the father-son team of Luis and Walter Alvarez, in conjunction with colleagues at Berkeley, to hypothesize in 1980 that a large asteroid crashed to earth about 66 million years ago, and that this collision resulted in the extinction that closed out the Mesozoic Era. According to this theory, all of the creatures in the wine-colored Mesozoic-aged strata whose fossils you have just collected were wiped out by this catastrophic impact. The piece of clay now in your hand, filled with stardust, is the impact layer, formed from the falling dust and debris thrown into the atmosphere at the time of the collision. According to this scenario, the Mesozoic Era ended with the biggest bang in history. Poison gas and uncontrolled wildfires spread through virtually all of the planet's forests; dinosaurs, scalded and burned, died by the millions; the upper 400 feet of the ocean became so acidic that the shells of marine life dissolved away. In this scenario half of all the species on earth perished quickly.

A crashing wave throws cold spray on your reverie and brings you back from Armageddon. The tide is rising, and if you don't hurry you'll be cut off by the rising sea, condemned to spend a cold night in this bay with its Mesozoic ghosts of animals and plants. You hurriedly pack your gear and glance at the rock that overlies the clay layer of the Cretaceous-Tertiary boundary. You see beautiful pink and white limestone, brilliant in the afternoon sun, rising above you for more than 100 feet, the earliest rock deposited in the Cenozoic Era. A swift hammer blow brings down a piece; with your hand lens you look at this shard and see once again larger planktonic microfossils, but species entirely different from those you found in the underlying Cretaceous strata. By the time the white limestone you now look at had been deposited, the slate of oceanic species, nearly wiped clean at the end of the Cretaceous, had been filled once again. And then, as has happened so often, just as you're about to take leave of this graveyard you see the smooth spiral shape of an old friend. You have no time to collect it, for the limestone that encases it is extremely hard and the tide is rising. You touch the shell of the nautiloid. Large and smooth, this Cenozoic nautiloid is the virtual double of the Jurassic specimens found on the beach at Lyme Regis and nearly identical to the living species you can catch in Fiji, New Caledonia, the Philippines, and a thousand other places in the Pacific. You have ten species of ammonites in your backpack, collected from the underlying Cretaceous strata. All died out before the clay layer at your feet was deposited. Only the nautiloid survived. And you wonder why.

How Did the Ammonites Die?

In the years since the Alvarezes proposed their impact hypothesis it has been the subject of unceasing debate—clamorous, emotional, often visceral debate. There is no middle ground to occupy on this issue: you are for it or against it. But even the most vociferous detractors admit that there is strong evidence that something of the sort took place some 66 million years ago, for virtually everywhere on earth where the Mesozoic-Cenozoic boundary is exposed there lies a clay layer that contains elements of the platinum group in concentrations virtually unknown elsewhere in the world. Platinum, a heavy element, is relatively common in the earth's interior but it's very rarely found on the surface of the earth, and then only in places where volcanic or tectonic upheaval has brought rocks from the earth's deep mantle to the surface. But there is another place where platinum can be found in some abundance: in outer space, among the wandering meteors. The class of meteors known as iron-rich meteors often contains substantial concentrations of platinum and its sister metal, iridium. It was the discovery of high concentrations of platinum and iridium at the first Mesozoic-Cenozoic boundary clay layer to be studied, at Gubbio in central Italy, that led the Alvarezes to propose their controversial hypothesis. A substantial part of the clay layer, they concluded, was made up of the remains of a huge meteor that had fragmented and virtually vaporized upon impact with the earth. By computer technology the Alvarezes even determined the size of this huge rock from space: it had to have had a minimum diameter of six miles to produce the amount of platinum-group elements found in the boundary layers.

As scientists scrambled to confirm or deny this hypothesis, new information from many Cretaceous-Tertiary boundary sites began to appear in various scientific journals. The exquisite sequence at Zumaya was soon sampled, as were many other sites both on land and in the sea. One of the early, astonishing findings was the discovery of shocked quartz grains within many of these clay layers. Such quartz grains had previously been seen at only

Cretaceous-Tertiary boundary exposed at Stevns Klint, Denmark (top), *and the clay layer containing iridium. The clay layer here is only a few inches thick (note the hammer below the clay layer for scale).*

two other kinds of sites: meteor impact craters and nuclear test sites. The presence of shocked quartz grains in the boundary clays soon became an important piece of evidence in support of the impact hypothesis.

In some scientists' view, these tiny bits of quartz crystal, usually smaller than a grain of sand, are even more certain indicators of meteor impact than the high concentrations of iridium and platinum. Glen Izett and Bruce Bohors of the United States Geological Survey, who collected bits of quartz from numerous boundary clay layers, demonstrated that the tiny crystals contained small internal fractures that could have resulted only from tremendous sudden pressure on the rock. According to these scientists, the collision of the gigantic meteor with the earth at the end of the Cretaceous disintegrated huge volumes of earth rock and threw much of it up into space—the so-called ejecta from the force of the impact. The tiny quartz grains now found in the boundary layer are the remains of this event, and the small shock fractures testify to the forces involved.

Could any other mechanism have produced the boundary layer with its excess iridium and shocked quartz? Some scientists think so. Geologists Charles Officer and Charles Drake (among others) have cogently argued that iridium and shocked quartz could be produced by volcanic action. According to this view, volcanic eruptions on a gigantic scale spewed rock material (including platinum and iridium) from deep within the earth into our planet's atmosphere 66 million years ago. The shocked quartz could have formed at the same time. This argument was strengthened by the finding that iridium is indeed being introduced into the lava and gas flowing out of Kilauea volcano in Hawaii. But critics of the volcanic hypothesis point out that the amount of iridium found in the Cretaceous boundary layers would have necessitated volcanic eruptions on an unimaginable scale. Although there is evidence of large-scale volcanic activity in India at the correct time, 66 million years ago, it doesn't seem to have been great enough to distribute iridium and shocked quartz all over the earth.

The physical evidence of the excess platinum-group elements and shocked quartz in the boundary clay layers strongly suggested that a gigantic meteor had hit the earth about 66 million years ago. But where was the crater? The collision of such a large body with the earth would leave a crater at least 100 miles in diameter. And did this impact, if it occurred at all, cause the extinctions observed at the end of the Cretaceous Period? Only the fossil record could answer that question. The ensuing controversy gave new life to the science of paleontology.

The killing mechanisms that would be triggered by such a huge strike on the earth can only be surmised, since (mercifully) there has never been an impact by anything approaching the size of the hypothesized end-Cretaceous meteor during recorded history. There have certainly been near misses, though. The most notable occurred in the spring of 1989, when an asteroid several miles in diameter passed between the earth and the moon. If such a body had hit us, there would have been tidal waves and earthquakes and perhaps worse. It has been proposed that the impact of a six-mile-diameter meteor would cause so much material to be thrown up into the sky that all sunlight would be shielded from the surface of the earth for three months or more. The earth, in this scenario, would have three months of night. Good-bye, plants. Even worse scenarios have been suggested. The passage of the meteor through the earth's atmosphere and the ensuing impact would release huge amounts of nitric acid into the oceans, in effect acidifying the upper 400 feet of the sea. Just as acid rain has made so many small lakes in the eastern United States and Canada sterile, the acidification of the upper layers of the ocean would kill off much of the life there—especially the plankton, on which so many creatures depend for food. But perhaps worst of all could have been the global wildfires. Analysis of the clay layers at various Mesozoic-Cenozoic boundaries reveal concentrations of soot. This organic carbon, discovered by E. Anders of the University of Chicago, could have been formed only by fire. And the concentrations he found were so high that Anders came to a startling conclusion: the amount of soot preserved in the Cretaceous-Tertiary boundary layers could have been formed only if more than 90 percent of all vegetation on earth had simultaneously caught fire. Imagine a view of the earth from space 66 million years ago if Anders is correct: all of the continents would be aflame, every forest and field ignited by the energy released by the huge asteroid's impact with the earth. And imagine the earth after such a conflagration: blackened landscapes, whole faunas exterminated by the fires, the only survivors those lucky enough to live deep in burrows or in freshwater lakes and ponds. Creeping from their deep sanctuaries, the survivors would see scenes from Dante: the blackened bones of the last triceratops herds, the skull of a tyrannosaurus that died in bellowing defiance as its burning flesh fell away from its bones.

There is no doubt that catastrophe gets good press. Global darkness, the oceans turned to toxic soup, the forests completely burned by the equivalent of a nuclear holocaust—such scenarios make good copy. But what about an alternative scenario? What if only a small meteor hit, perhaps no more than a half mile in diameter, large enough to produce the iridium- and platinum-rich clay layers found at many Mesozoic-Cenozoic boundary sites but not large enough to have the dire consequences of impact by a larger meteor? And even more fundamental—could any meteor, regardless of its size, produce extinctions of the magnitude found at the end of the Mesozoic, everywhere in the world? Wouldn't even the largest meteor extinguish life only in the region where it hit? What if the impact was small but came at the end of a protracted period—say two to five million years—of mass extinctions caused by entirely different factors? This is the view of many respected earth scientists.

How to choose between these two views? Surely the best tests of the competing hypotheses would involve analysis of the fossil record of the latest Cretaceous period, the time interval immediately preceding the extinction. If long-term processes, such as a change in sea level or global cooling, produced the extinctions at the end of the Mesozoic, the various animal and plant groups should decline gradually in numbers and diversity before they finally died out. If, on the other hand, the various groups were driven to extinction by a sudden catastrophe, the fossil record would show constant numbers and diversity of species right up to the end of the Mesozoic, followed by a virtually instantaneous decrease or elimination of entire species. Clearly, one or the other of these two alternatives should be apparent when the fossil record is carefully analyzed. Or so it seemed in the early 1980s. And one of the best groups available to test these two alternatives were the ammonites. Ammonites did not survive the end of the Mesozoic. But did they die out suddenly or gradually? The investigation thus became a high-stakes game, the results eagerly anticipated by both pro- and anti-impact camps. And the cliffs of Zumaya became one of the best-known laboratories for conducting this investigation.

The Extinction of the Ammonites

Until my first glimpse of Boundary Bay in 1982, I had been little concerned with the controversy about the cause and consequences of the Mesozoic extinctions. I had mostly been studying the biology of the nautilus, the last remaining chambered cephalopod and member of one of the select few lineages of animals that survived the end of the Cretaceous, the period that ended the Mesozoic Era. In June 1982 I had journeyed to a beautiful French marine laboratory in Banyuls-sur-Mer, a small town on the Mediterranean, to study Sepia, a cuttlefish that has a buoyancy system similar to that used by the nautiluses, and by the extinct ammonites as well. This visit was the start of a six-month sabbatical, my reward for six long years of teaching. Ammonites and extinctions were far from my mind that June. But I soon found that ammonites were much on the mind of my host, Sigurd von Boletzky, a cephalopod specialist working on the growth and rearing of sepias, among other creatures. Sigurd had been conducting elegant, long-term experiments on newly hatched sepias from the Mediterranean, and had found that their captivity led to a marked stunting and dwarfing of shell and tissue growth. Since the stunted growth was recorded in the shell (and thus could potentially be observed in fossil forms), his work had caught the attention of one of the great experts on Mesozoic ammonites, Jost Wiedmann of Tübingen University, in what was then West Germany. Wiedmann had spent most of his career studying the history of Cretaceous ammonites. One of his major interests was in how and especially why they had become extinct. How could such a diverse group disappear from the earth after so long and successful a reign? Wiedmann sought answers both in the rock record and in the biology of living cephalopods. He had long been investigating the biostratigraphy of northern Spain, where in the 1960s one of his colleagues had discovered the late-Cretaceous strata at Zumaya. So Wiedmann traveled there and made an astonishing discovery: the shells of the ammonites he found in the highest (and thus youngest) Cretaceous exposures of the Zumaya coast were, in his opinion, abnormal. To Wiedmann they appeared to be stunted dwarfs; they looked almost exactly like the cuttlefish he was raising in captivity. Wiedmann thought he had the first major clue to the death of the ammonites: the information from Zumaya suggested that some aspect of feeding or food supply became abnormal near the end of the Cretaceous. Here was evidence that at least for this group of animals the end came gradually, not suddenly, and was related to long-term changes in the food chain. The next question became: What happened to the food chain? Wiedmann had a ready answer: It was disturbed by a rapid change in sea level.

During my graduate school days, in the early 1970s, paleontologists taught that the Mesozoic Era came to a gradual end, and that the extinctions had been caused by earthbound events. The leading culprit was thought to be a rapid change in sea level. It has long been known that the level of the sea rises and falls with some regularity; indeed, because of the changes in the volumes of ice brought about by the Pleistocene glaciations of the last 2 million years, the level of the sea has varied as much as 400 feet. Even as recently as 15,000 years ago, at a time of maximum glaciation, the level of the seas was about 350 feet lower than it is now. The seas have been rising ever since, and if we may judge by global warming currently under way, they should continue to rise. But rapid changes in sea level, dramatic as they are, are rare events in the history of the earth. Changes in sea level more commonly are gradual and are caused by changes in tectonic processes rather than in the ratio of seawater to ice. The level of the sea has risen and dropped many times, but usually very slowly.

The great scientific revolution of the 1960s, which showed us that the surface of the earth is composed of huge, thin plates that ride piggy-back on moving slabs of hot, viscous magma, yields an explanation for long-term changes in sea level. The earth's heat engine runs the motors of plate tectonics; depending on the amount of heat rising from the interior of the earth at a given time, the great convection cells of magma that cause the plates to move can increase or decrease. When heat flow increases, the whole mechanism runs faster. A by-product of this higher rate of convection is a slight decrease in the volumes of the oceans. The mid-ocean ridges, where new magma wells up to form new ocean floor, enlarge somewhat when heat flow is high. Enlargement of the ridges in turn slightly reduces the volume of space in the ocean basins, causing seawater to lap onto the continental lowlands. When heat flow is reduced, the mid-ocean ridges subside slightly, thereby enlarging the ocean basins, and the seas flow off the continental edges.

The Cretaceous Period had been a time of very high heat flow, and consequently a time when much of the continental lowlands were covered with shallow inland seas. The entire central portion of the United States, for instance, was covered by such a sea, and so was much of Africa, South America, Asia, and Europe. Near the end of the Cretaceous, however, heat flow apparently subsided, causing a rather rapid withdrawal of the seas from the land. As the seas flowed from the land after so many millions of years, entire ecosystems were disrupted both in the sea and on land. The conjunction of one of the most rapid drops in sea level known in the history of the earth with one of the largest extinctions revealed by the rock record was, in most scientists' minds, more than simple coincidence. It was certainly more than coincidence in the mind of Jost Wiedmann. He viewed the death of the ammonites as a direct result of the large-scale change in sea level at the end of the Mesozoic Era. And he had good reason to come to this conclusion: earlier in his career Wiedmann had demonstrated that the number of ammonite species known from the fossil record over the various time intervals of their Paleozoic and Mesozoic existence seemed to be related to sea level. During times of dropping sea level, the number of ammonites diminished; when the sea level rose, the ammonites proliferated. It thus seemed logical that the final decline and extinction of this long-lived group occurred during the most sustained and dramatic drop in sea level known during the Mesozoic Era. Wiedmann reasoned that the dropping sea level disrupted the ammonites' food supply.

Boletsky's demonstration that sepias became stunted when their food supplies changed, coupled with Wiedmann's discovery of dwarfed ammonites at the highest levels of Zumaya, seemed to make the argument irrefutable. Only slight doubts emerged. Why weren't all of the cephalopods eradicated? That our seas contain such wondrous creatures still as the octopus, the squid, and, most hoary of all, the nautilus indicates that not all cephalopods were snared by the falling seas. How had these others survived when the seemingly indestructible ammonites fell to dust? It is clear that the ammonites' diversity was indeed influenced by the rise and fall of the seas. But such changes had occurred often during the long history of the group. A change in sea level was probably not the culprit in the extinction of the ammonites.

The Ammonite Gap

Jost Wiedmann was with me during my first journey to Zumaya, in the summer of 1982. I was amazed by the thickness of the outcrops, by the virtually continuous exposures, and most of all by the dramatic difference between the highest Cretaceous rocks and those of the lowest Tertiary. We spent two days on the Zumaya outcrops, collecting an occasional ammonite fossil, and at the end of our second afternoon together we headed in different directions: I took a train back to my marine lab in France and Wiedmann continued on his way along the Spanish coast toward other outcrops and other problems, for to his mind the nature and cause of the ammonites' extinction were already explained. I was left with a nagging uneasiness, however. First, even though I looked for several hours, I was unable to find any ammonites in the Boundary Bay exposures; the youngest 30 feet of Cretaceous exposures at Zumaya seemed to be empty of them. And second, although most of the ammonites I did find lower in the Zumaya exposures were indeed small, and thus looked like the dwarfs that Wiedmann described, I was able to collect several larger specimens as well, and they looked normal enough to me.

My first visit to Zumaya was thus essentially a sightseeing trip. I had no intention of continuing work there, for the problem seemed finished. So when I visited Wiedmann at his Tübingen laboratory, I was amazed to find that he too had never found an ammonite fossil anywhere near the Mesozoic-Cenozoic boundary.

Thus started a long collaboration. I returned to Zumaya for several weeks that summer, and then again in 1984, 1987, 1988, and twice in 1989. As the number of ammonite fossils swelled with each collecting trip, a detailed picture of the final history of this group began to emerge. It quickly became apparent that the numerous "dwarf" ammonites were, in reality, perfectly normal juveniles, whose small size was due only to their immaturity. Ammonites—like the still-living nautiluses—left a distinct morphological clue about their growth. Nautiluses (and apparently the ammonites as well) were not creatures that grew throughout their lives. Like humans, they reached a certain adult size and then quit growing. The slowing of growth immediately preceding the final adult size is marked by changes in the spacing of the last two or three septa formed within the shell and by changes in the shape of the outer shell wall. The small Zumaya ammonites showed none of these changes. They therefore could not have been dwarfed adults; they were not adults at all. The sea bottom at Zumaya was apparently a breading ground for ammonites, a place where the young lived and grew and died in prodigious numbers. But adults could also be found, including many very large ones.

The other astonishing finding was the scarcity of ammonites just beneath the boundary. The Zumaya exposures are about 800 feet thick and spread out over a mile of rugged coastline. The goal of our research project was to document the record of ammonites during the time these strata were deposited, thus recording the last 2 million years of ammonite history, by specimens from the entire 800-foot section. But the most important question of all concerned the final fate of the ammonites—specifically, determining the level at which the final specimens could be found. We wanted to know whether the last ammonites were killed off in the catastrophe that ended the Mesozoic, and how close to the clay layer the last ammonite could be found. We could find out only by spending many hours searching the strata in Boundary Bay.

As the tide ebbed we often worked away on the cliffs leading to Boundary Bay until the water had dropped so far that we could scramble up over the last rocky barrier and drop down into the bay itself. And then, in that quiet bay, I scanned the rocks for ammonites in the limestone and shale of the last Cretaceous strata. Sometimes I would quarry, using hammer and chisel to bring down rocks from the overhangs, praying that the hard hat on my head would deflect any stray blocks that slid off the cornices jutting out above me. Over the years the numbers of ammonites mounted under such investigative onslaught, until a half-dozen species and many individuals were documented from Boundary Bay—but never from the last 30 feet of the Cretaceous. It was as if a 30-foot-wide dead zone preceded the final catastrophe, suggesting that ammonites died out well before the end of the Cretaceous Period. But finally, on a rainy day, I found a fragment of ammonite within inches of the clay layer marking the boundary. Slowly, over the years, several more were found in the highest levels of Cretaceous strata at Zumaya. Ammonites appeared to have been present for Armageddon after all.

Over the years the rocky cliffs of Zumaya became like friends, and it was easy to forget that this land was in political turmoil. The weather on the Bay of Biscay is very changeable, and a day that started out in warm sunshine could often finish in rain. But many days were glorious, and we geologists must have seemed incongruous among the gay summer bathers, decked out as we were in the paraphernalia of field geology. No doubt we attracted comment among the local Basques, but they largely ignored us, for we seemed to be harmless eccentrics, wasting beautiful afternoons staring at the rock surfaces or scaling the cliffs like human flies. But in 1984 we must have seemed far less harmless, for that spring I brought a gasoline-powered rock drill to Zumaya. With that drill I could remove small cores of rock, which I then could analyze for magnetostratigraphy, a procedure that allows fine-scale correlation of rock strata by measuring the magnetic properties of their constituent minerals. The drill was modified from a chain saw, and like those infernal machines, it produced an ear-splitting racket. Coring of the rocks was a slow, exacting business, and the constant malfunction of the various pieces of equipment drove my small field party nearly mad. But the drill was having an even worse effect on the local people. On a gray afternoon I was drilling cores in Boundary Bay when a glint of color caught my eye. Looking up, I was stunned to see a dozen machine guns pointed at me, held by Spanish soldiers in blue cocked hats. One of them came down and barked questions in such rapid Spanish that I couldn't understand him. The soldiers were understandably nervous about loud noises; the week before, several members of the local constabulary had been ambushed and gunned down by Basque separatists. The army was making sure I wasn't one of them, getting ready to dynamite the cliffs of Zumaya. After searching all of my bags and finding nothing but fossils, they left in sullen silence.

Three summers later the other side paid me a visit. On a hot, beautiful August afternoon I was sitting in Boundary Bay, smacking a ten-pound crack hammer against a rocky outcrop, when a shadow fell across me. Looking up, I saw a tall man standing over me. I smiled, expected to hear a question about what I was doing so lustily on this fine day. But I was met by a stony stare and a charge that I was despoiling the landscape. I looked guiltily at the pile of rubble I had accumulated and tried to explain that erosion from the sea broke off far more rock each year than my puny efforts. But he pointed to one of the nearby core holes, an inch wide and two inches deep, visible (but eroding) evidence of my coring efforts of three years before. Then he pointed at me. I prevaricated, and the man directly accused me: "You drilled these holes. We know. We have been watching you these past years." With that he strode away. And soon after I did too, and I stayed away for two years, waiting for tides and memories to smooth away the industry of paleontology. By then I had discovered new evidence about the extinction of the ammonites, in rocks not far from Zumaya.

The Bay of Biscay

Zumaya sits on the Bay of Biscay, a region of stunning rocky coastline. It was my hope that somewhere along that coast I would find other sites like Zumaya. A recurring dream of the paleontologist is to find an entirely new locality to explore and collect, a place where other geologists or amateur fossil hunters have not already been at work. The geological maps told us that rocks of the same age as those at Zumaya could be found both to the east and to the west; and in hope of finding other Cretaceous-Tertiary boundary sections that could yield new information about the extinctions, in 1987 I began to explore the Biscay coast. In the course of traveling a hundred miles in each direction from Zumaya I found two new places rich in ammonite fossils: cliffs at Hendaye, a beautiful small town right on the Spanish-French border, and a second site about ten miles up the coast in France, near Biarritz. Not that I was the first to identify these places as Cretaceous-Tertiary boundary sections, but neither of them was thought to contain more than a few ammonites. Yet each of these sections proved to be a treasurehouse of ammonites, so many and in such diversity that the sites easily rivaled Zumaya. And even a greater surprise was the finding that, unlike Zumaya, where hundreds of hours of searching in the highest portions of the Cretaceous had yielded but a single ammonite species, Hendaye and Biarritz yielded numerous ammonites close to the Cretaceous rock, just beneath the clay layer that marks the boundary.

In appearance the Hendaye and Biarritz cliff sections were strikingly different, but each had a haunting beauty of its own. At Hendaye a rocky headland juts out into the sea. The entire region is a large parkland, dominated by a castle at its highest point. You enter on a long sandy beach and then climb into the hills, past pastures filled with sheep. As you walk across vast fields you have a spectacular view of the sea and when you finally reach the rocky strata, you work in complete solitude. The large park grounds are also, unfortunately, dotted with relics of human combat: huge concrete bunkers and pillboxes sprawled among the greenery, part of Hitler's Atlantic wall, now covered with vines and vegetation.

The crucial boundary section at Hendaye is beautifully exposed on a rocky cliff near the sea. It's a lonely place, reached by a hair-raising slide down a crumbling talus slope. The boundary between the Cretaceous and Tertiary eras is found within a large cave, for the clay layer is softer than the underlying or overlying strata, and the constant pounding of the sea has hollowed out an opening large enough for a person to crawl through. Here too, as at Zumaya, the tides dictate the work schedule. Sometimes I would arrive too early and wait impatiently for the sea to drop. My first visit to this section was a revelation. At first I was dismayed, for I saw the telltale signs of hammer and chisel, gouges made in the Cretaceous rock by earlier fossil hunters. But I soon found that the rocks around me had been mined not for ammonites but

Thinly bedded Cretaceous strata exposed on the beach at Hendaye at low tide. The fossil-collecting sites and Cretaceous-Tertiary boundary are seen in the distance.

The Cretaceous-Tertiary boundary at Hendaye, separating the underlying Cretaceous shales from the overlying, more thickly bedded Tertiary limestones, is visible as a dark line running from upper left to lower right.

for the large crystals of fool's gold found just beneath the clay layer. These large nodules of pyrite could themselves be evidence of catastrophe, for such crystals grow most frequently in sites of low oxygen and rotting animal and plant matter. After my experiences at Zumaya, where years of searching yielded only the slightest evidence of animal life near the Cretaceous-Tertiary boundary, I was overjoyed to find a score of ammonites within the last meter of Cretaceous rock during my first hour at Hendaye. In fact, I found more ammonites here than any other kind of fossils. Gradually the number of ammonites I found in the last meter of the Cretaceous at Hendaye mounted, adding wonderful new information about the last intervals of an era.

In contrast to the stark isolation of the Hendaye site, the Cretaceous-Tertiary boundary section on the seashore south of Biarritz was like a carnival spot. I found it behind a breathtaking expanse of sandy beach, after making my way through multitudes of summer bathers. The long, golden beaches in this part of southern France are popular vacation spots, and I soon found that to do my work I had to come either early in the morning or late in the afternoon, after the sun worshipers had abandoned the beach for the local cafés; during the long, hot afternoons the critical sedimentary rocks were invariably covered by drying surfboards, towels, and human bodies. The Biarritz section presented unforeseen occupational difficulties, for the stretch of sand immediately in front of the Cretaceous-Tertiary boundary is a nude beach that the gay community has claimed as its territory. Biarritz is the San Francisco of France. We geologists were always the only denizens of that beach who sported any clothes at all, necessary protection against the sun and the rock chips that our hammers and chisels sent flying. We seemed to represent a challenge to the other denizens of the beach, and we learned to cherish the cloudy days when we could work in solitude. As elsewhere along the Bay of Biscay, the contact between the Cretaceous and Tertiary periods in the Biarritz region was marked by a clay layer, and within inches of it we eventually found many ammonites.

A decade of study in the Bay of Biscay region slowly brought together a picture of the final few million years of the Age of Ammonites and gave stark testimony to the end of this long-lived group. At Zumaya few ammonites apparently existed on or above the seabed that is now preserved as coastal rocks. I now think I know why: at the end of the Cretaceous, Zumaya was in the deepest part of the basin, at depths too great to sustain many ammonites. The thickness of the seabed at Hendaye and Biarritz seems to indicate that these sites were in shallower water.

At least ten and perhaps a dozen species of ammonites lived in the Bay of Biscay right to the end of the Cretaceous Period. Most of them were streamlined swimmers or forms that lived in the mid-water regions, relatively safe from the predators that lurked far below them at the bottom and from those on the surface far above. All of these ammonite species could live at depths much below the shallow waters where the vast majority of ammonites lived before the final few million years of the Cretaceous Period. By 70 million years ago ammonites were obsolete at the bottoms of shallow seas, and if they wandered into such places, they must soon have became meals for the shell-breaking carnivores. And then they were swept away to final extinction 66 million years ago, leaving only one close relative: the nautiloids.

The Survival of the Nautiloids

Perhaps it was luck that helped the nautiloids survive; somebody wins a lottery every day. But I suspect a more likely explanation. Much of the original impetus to study the nautilus came from a desire to know more about its reproductive system. Why does each female nautilus produce only a dozen large eggs each year, when other living cephalopods produce thousands, even tens of thousands? The nautilus eggs seem to be laid and kept at very great depths in the sea—perhaps as much as 300 to 1000 feet—during the year it takes them to develop. I suspect that when the great catastrophe killed off all of the juvenile and adult ammonites at the end of the Cretaceous, it destroyed all of the nautiloids as well—except for their slowly developing eggs, preserved by their great depth. We can imagine the postcatastrophe world of 66 million years ago, the dead creatures piled high on the beaches, the seas emptied of all those creatures that had lived in and on the plankton—ammonite hatchlings, for instance, which probably floated on the surface for some time before, in happier times, they finally joined their parents down below. Perhaps some of the larger ammonites escaped the catastrophe for a time, also protected by depth, only to wither and die when the food chain was broken by the death of the plankton. There is no controversy on that score; the end of the Cretaceous Period witnessed the greatest calamity ever to visit the floating creatures that make up the sea's plankton. It may not have been the larger ammonites that perished but their progeny; the nautiloids may have survived their ammonite descendants because of their radically different reproductive styles.

They live still, the nautiloids, represented by one genus and five species, in the remote depths of the Pacific Ocean. By 1987 I had a single chance left to visit these survivors.

The Nautilus of Vanuatu

Seattle, Washington, December 6, 11:00 P.M.

I'm driving home from my university office, weary from last-day exertions; tomorrow I leave for the Republic of Vanuatu, once called the New Hebrides, west of Fiji. I've spent the day gathering, packing, and taking care of the emergencies that always seem to crop up just before a departure. My mind dwells on bills paid and unpaid, letters that should have been answered, good-byes forgotten. Outside, a cold, blustery wind lashes freezing rain against my car. I pulled into my driveway and unload the last paraphernalia yet to be packed: batteries, film, a new face mask, an X-ray machine. As I walk toward my house with my arms full, a light catches my eye. The strong south wind of the Seattle winter has briefly pushed the clouds from the night sky. A half moon is just now rising, washing out the glory of Orion. Below Orion twinkles Sirius, low in the south. I look back at the waning half moon. And finally, so late, I realize where I am going. At the end of my long voyage Sirius will no longer be low in the southern sky; it will be shining overhead. And it will be not the brightest star in the sky but a lonely second to Alpha Centauri. Orion will be on the meridian, and upside down to my northern senses. But the most important thing will be the moon. Far to the west and south, the moon has not yet risen. The sun is only now setting, and within an hour the black tropical night will have fallen. For many hours no moonlight will be filtering in the seas to imperil the nocturnal voyagers in front of the coral reefs. I stare at the pale moon, about to be obscured once again by the winter clouds of Seattle, and know that at this moment the nautiluses of Vanuatu and New Caledonia and Palau and countless other islands are just now beginning the thousand-foot voyages that will take them to the tops of the reefs, with no moon to betray them. And so it will be for the next two weeks, until the waxing moon once again will inhibit the vertical migrations of the nautilus. I'm smiling to myself now; life has taken on a nice sparkle. There's something wonderful about knowing that the phase of the moon will be one of the most important things to affect your life over the next few weeks.

North Effate Island, Vanuatu, December 22, 5:30 A.M.

I'm awakened by the pale light of the tropical dawn, telling me that my last day in the Republic of Vanuatu has begun. I roll over on my mat to see that Andrew, my assistant from the Vanuatu Department of Fisheries, has already risen and is staring at the sea. The wind is already beginning to ruffle the sea's surface: a promise of waves for the day. My eyes are scratchy from last night's seawater. I made my last dive the night before, looking in vain for nautiluses in the shallow waters in front of the village. For a week the nights were dark, and I saw the silent nautiluses in water as shallow as three feet deep; but last night I dived in the silver glow of moonlight. Growing brighter nightly, the moon now poses a danger to the shallow-water nautiluses. They won't come into the shallows for the next two weeks, until the treacherous moon once more loses its power over the night. My mind caresses the last two weeks while Andrew heats water for our coffee. For once I'm in good shape for a last day, for I have largely completed my mission here. Seventeen nautiluses are in the freezer, ready to be taken back to laboratories in the United States. Study of their DNA will tell us much about the ancestry of this creature. Andrew and I share a bowl of rice left over from the night before. Outside, daylight is steadily increasing, and soon the fierce sun will hurtle up into the sky. My last goal is to bring any nautilus caught today back to America—alive.

On the Road, 6:00 A.M

The countryside of Vanuatu is a tropical paradise. The flamboyants are in bloom, splashes of red against the lush foliage of the rain forest. We're driving through this countryside in our flatbed truck, going the short distance from the village where we stay to the fisheries anchorage. Andrew drives the truck at terrifying speed, defying gravity and potholes on the American-built World War II–vintage dirt road that's the principal highway around the island of Effate. We've been fishing the north end of the island, two hours by this road from Port Vila, the largest town and capital of Vanuatu.

Vanuatu has been independent since 1981, when the government administered jointly by Britain and France hauled down their flags for the last time. The transition from the colonial New Hebrides to the independent Vanuatu did not go smoothly. I was in nearby New Caledonia at the time, and watched refugee boats arrive with shiploads of French colonists, possessions in hand, fleeing from the newly established republic: a replay of scenes enacted in earlier decades in Algeria and Vietnam. Since then, Vanuatu had been a bee in the collective bonnet of the South Pacific. Inspired by their canny prime minister, Walter Lini, the Vanuatuans had cleverly played the three major powers of the region—France, Australia, and the United States—off against each other, to their own great advantage. By granting the Soviet Union unprecedented fishing rights and establishing communication and exchanges with Libya, Vanuatu received attention to its views from normally insensitive Western powers. My presence in Vanuatu was unprecedented as well. Vanuatu had in the recent past turned down requests for research privileges by American marine biologists. It has taken me a year to receive permission to study the local nautiluses.

And what extraordinary nautiluses the Vanuatuans had. I had seen the shells some years previously, and knew immediately that these nautiluses would be worthy of study. Nautilus shells from Vanuatu look very much like the shells of the species *Nautilus macromphalus* from New Caledonia, with their abundance of rich purplish-red bands. But the shells from Vanuatu contain in the umbilical shell region the plug of calcium carbonate that characterizes the species *N. pompilius*. Long ago, perhaps during the Great Ice Age of the last million years, when the sea level was as much as 400 feet lower than it is now, the nautiluses of New Caledonia and Vanuatu may have belonged to the same gene pool. As the great ice sheets melted, the sea began to rise, and in the increasing expanse of water these two populations must have become isolated. In at least one other respect the nautiluses of these two island groups appear to be closely related: only in New Caledonia and Vanuatu do they come all the way to the surface of the sea at night.

I had historical motives as well to study the Vanuatu nautilus. It was from the New Hebrides that the nineteenth-century anatomist Sir Richard Owen procured the first nautilus soft parts and for the first time described the anatomy of this living fossil. His extraordinary report on his findings alerted Western scientists to the importance of the nautilus in their efforts to understand the biology of modern as well as extinct cephalopods. But in all the century and a half since Owen's work, no scientist had studied even one other individual of this elegant species.

Fisheries Pier, 6:30 A.M.

In the company of our captain and his one-man crew, Andrew and I row out to our boat, the *Etelis*, already pitching uneasily at its anchorage in the gusting wind. We start the diesel motor of our ship and slip into the wide bay with the wind at our backs, the large boat riding smoothly in the trailing sea. We all know, however, that we will pay dearly on the return trip.

Aboard the Fisheries Ship *Etelis*, 8 A.M.

We're far out from land and in sight of the bright-red buoys that mark the position of our nautilus traps. In the distance the steep cones of volcanoes are visible. Our traps, giant cubes of iron bars and metal screen, have been in the water since the previous Friday, baited with skipjack and shark. The traps were trailing thousands of feet of line behind them when we shoved them over the side of the boat into deep water several days ago. With difficulty we snag the first buoy and attach the line to the hydraulic winch. The wind blows the boat around as we begin to raise the trap from its resting place a thousand feet below us. It takes us fifteen minutes to haul it to the surface. I watch its ascent on the fish finder; as the trap makes its way upward, another shape lifts off the bottom and follows it. This chase is recorded as two streaks of carbon on the paper of the echo sounder. The second is undoubtedly made by some large deep-water shark in frustrated pursuit of the rapidly rising trap. The chase ends a hundred feet below us: one carbon streak, the trap, gliding ever upward; the second, confronted perhaps for the first time by bright sunlight, turning back and down to the eternal night of the abyss below us. We peer anxiously over the side. Streams of bubbles break the surface. Now we know that the trap isn't empty, for the bubbles can be coming only from fish caught in it, their swim bladders bursting with the gas that has expanded as the water pressure has dropped during their rapid ascent. The trap breaks the surface, crashing against the side of the boat in the heavy sea. The four of us strain to pull it into the boat. Andrew exhorts us, and with a cry of "heave!" we finally haul the trap on deck. It's full of writhing bodies: a giant gray eel with slashing teeth; two large, brilliantly pink deep-water snappers with swim bladders bloated outward from their mouths; an octopus, the most feared predator of the nautilus, killing its foes with cleverly drilled holes; a giant spiny crab, massive claws held menacingly; flapping shrimp, beautifully colored with red racing lines; and two nautiluses. The nautiluses are among the most beautiful on earth, colored with rich magenta stripes. We turn about into the wind and head for the second buoy.

Fisheries Pier, 1:00 P.M.

We arrive back at the anchorage with our five large traps covering the boat. One of the traps is horribly deformed. It snagged on some deep ledge or promontory, and we had to maneuver for an agonizing hour before we could finally winch it up. With difficulty at our rocking anchorage we transfer the traps to a smaller boat that will take us to shore. This is normally siestatime, for the summer sun beats down with fierce strength; sweat is streaming from us as we manhandle the awkward traps onto the flatbed truck. The rolling sea has left us with unsteady legs, and the sand seems to sway beneath our feet. I've captured four nautiluses in all. We transfer them into cooled seawater and load them onto the truck. My small bag is already packed, and I look around one last time at this place, one more place that has been a home I'll never see again.

Along the Effate Coast, 2:00 P.M.

We're sitting in a small village by the sea, partway back to Port Vila. The rough road has sloshed all of the water out of the bucket holding my nautiluses. I'm waiting for newly acquired seawater to cool, using the last of my precious seawater ice to drop the temperature of this tepid lagoon water down from its lethal level. There is no escape from the heat or the flies. I'm covered with flies as I transfer the nautiluses to their new seawater. The places of the flies I brush off are promptly occupied by their friends. Some burrow into my hair; others cake my legs, drawn to the cuts that all of us received from the jagged wire of the traps. Andrew and I must drink fluids constantly to make up for the perspiration streaming from us. We finally resume our journey, racing against elevated temperature and oxygen starvation for the nautiluses.

Port Vila, 4:00 P.M.

The four nautiluses are swimming about in the large tank in the laboratory of the Port Vila Fisheries center. I am surrounded by mosquito coils, for two strains of malaria are common in Vanuatu, and mosquitos abound after the nightly rains. The heat and humidity are stifling. I have much to do before my 3:30 A.M. departure.

Port Vila, 5:00 P.M.

I'm barreling through the main street of Port Vila in the fisheries' Land Rover, looking for rubber bands. Nautiluses can be transported live if they're sealed in bags of cooled seawater and pure oxygen. I have the large plastic sacks I need, but in all of Port Vila I can't find large rubber bands to hold them closed. This is unbelievable. Why do I always forget something crucial on these trips? Beside me sits the large cooler that will hold the nautiluses in their sacks and the heavy-duty Thermos jug to transport the frozen pieces of nautilus flesh destined for genetic study in the United States. But where can I find rubber bands?

Fisheries Base, Port Vila, 7:00 P.M.

I'm engaged in murder at the moment. Soon after my arrival in the afternoon I took one of the living nautiluses and put it in the freezer. The intense cold would softly send the nautilus to its death. I'm exhausted. A sack of mangoes lies before me, my restorative. My instruments are prepared, waiting for the victim, the passage from life to numbers. Like the buffalo of the American Indians, this nautilus will be thoroughly used. I've come to terms with scientific murder; I can rationalize this death. The nautilus population in the waters around Effate alone must number at least in the tens of thousands. But the fact is that I've killed sixteen animals that were born about the time Sergeant Pepper taught the band to play. I don't like the killing. With this trip my voyages to the Pacific are over. I'll miss the trips, but not the killing.

The nautilus feels like a balloon filled with cold water in my hands, and I marvel once again at the compact anatomy. I turn the animal over to expose its underside and gently peel back the skirtlike mantle tissue to expose the four gills. Following the stem of the bladelike gills to their source, I see the four sacs of the kidneys filled with white. With sharp scissors I open the kidneys, and a paste of granular crystals emerges. Another marvel of the nautilus is carefully weighed and put into tiny bottles: kidney stones, each less than a millimeter in diameter, present in the tens of thousands in each kidney sac. A large adult nautilus can have up to three or four grams of these microscopic rods and spheres of calcium phosphate. It's not yet known if all this calcium phosphate plays some role in shell growth or is simply accumulated waste.

The reproductive system is next. Unlike so many other mollusks, which broadcast their gametes into the sea, nautiluses must copulate to ensure fertilization. Their many arms interlocking, the male and female embrace for hours, until a packet of sperm is transferred from the male to the female. The great Victorian zoologist Arthur Willey thought that a nautilus couldn't reproduce until its shell had attained its full growth, but he was never able to prove his supposition. That proof awaited the brilliant work of Desmond Collins of the Royal Ontario Museum. Collins's research finally revealed the timing and nature of the maturation process; but the work was done on *Nautilus macromphalus*, the species found in New Caledonia. The Vanuatu nautiluses could confirm whether or not Collins's findings extend to other species as well. The specimen I'm dissecting was chosen for death over its three fellow captives precisely because it was the only one of the three in the crucial phase of final growth.

Fisheries Base, Port Vila, 10:30 P.M.

I've worked my way to the alimentary canal. All during my slow dissection I've been glancing at the bulging crop and wondered at its contents. The trap that caught this nautilus was baited with fish; anything present in the nautilus's crop, stomach, or intestines other than fish will be natural food. I carefully cut into the gut and jump back as fresh crustacean carapace material comes spurting out. Spines, claws, pieces of abdomen—all show the gaudy black-and-yellow stripes of the common reef lobster of Vanuatu. But this nautilus was captured at 1000 feet, about 990 feet deeper than the habitat of the lobster. A few nights ago I collected two nautiluses in 20 feet of water. The gut contents of my dissected specimen confirm a generalization suggested by those earlier sightings: nautiluses in Vanuatu undertake great vertical migrations from deep to shallow water at night, and then return to the safe deeps with first daylight, to cheat the shell-breaking predators.

Fisheries Base, Port Vila, 11:45 P.M.

I'm beginning to panic. I have yet to prepare my three remaining nautiluses for live transport. The lab around me is a wreck. I must still wash and pack my diving gear and gather my scattered scientific paraphernalia.

Fisheries Base, Port Vila, 1:30 A.M.

Somewhere in my brain a demon is whispering. Perhaps the nice Australian voice reconfirming my flight said that the flight leaves at 2:30 rather than 3:30 A.M. The more I think about it, the surer I am that the departure time of the flight has been changed. It's time to pack the nautiluses. I move into the large open-air shop, searching in the dim light for the oxygen and acetylene bottles I saw there earlier. I partially fill a large plastic bag with cooled seawater and put one live nautilus in. The gallon bag has just enough water to cover the nautilus. I close the top of the bag around a hose connected to the oxygen bottle and turn on the gas. Pure oxygen fills the sack until it expands into a giant balloon. With fumbling fingers I pull the oxygen hose from the bag and try to close off the top of the bag before any of the precious oxygen can escape. I take bailing wire from the coil on the floor and with a pair of strong pliers wire the end of the plastic bag closed. I put the bag into the cooler and turn to the second nautilus, conscious of time now, hurrying. As I close the second bag, the sharp end of the bailing wire shreds my hand. When the third bag is closed, a large gout of my blood remains inside it. As the bag turns pink, the nautilus goes crazy, swimming madly about with its tentacles extended in a maximum search posture, hunting for the source of this rich food scent, mammal meat that must be nearby. Fair enough, I think. I've eaten your relatives, after all.

I have no more iced seawater and no more time. This nautilus will know me well after this voyage in the pink seawater. This last bag is packed, the cooler closed. I drag the heavy cooler to the waiting Land Rover, where my other bags are already stowed. I take a look around, a last moment of peace. The bright southern stars in their unfamiliar constellations are scattered over my head. My hand is throbbing, another scar to add to the collection gathered on this trip. A faint knocking comes from the inside of the cooler: the nautiluses swimming in their captivity, perhaps glorying in the rich, oxygenated seawater that will be their home for the thirty hours of my return voyage. It's time to go. I cinch my belt tighter to keep the ragged cuffs of my baggy trousers out of the mud. Only three weeks ago I was about to throw this belt away because it was too small. Plenty of room now. I think back to my last night in Seattle before this trip and to what I have learned on this voyage. A more insistent knocking from the cooler sets me going. I have an hour until departure and airline counter clerks yet to bamboozle: I have three times the maximum allowable baggage weight.

As I head for the airport I can't know that these nautiluses will survive their journey and be the subjects of both a nationwide

An adult nautilus.

television report and a United Press International newspaper story that will be reprinted in hundreds of papers in the United States and Europe. One of the three nautiluses bumping around in my cooler will be photographed in its new home in the Seattle Aquarium, but the newspapers will mistakenly print its picture upside down, like the constellation of Orion over my head as I drag my bags into the Vanuatu airport. The waxing half moon has already set, letting the nautiluses on the deep reefs around me finally begin their upward journeys. Each night now the moon grows brighter and fills up more of the night as it approaches fullness. But for both the captive nautiluses and me, the phase of the moon is no longer of any importance. We're going home.

Ammonites as Living Fossils

The ammonites are long dead. Of all the old forms with external shells, the nautiloids and ammonites, only the nautilus remains. For that reason alone the ammonites may seem odd creatures to talk about in the company of the long-lived survivors that still exist on our planet. But I class them with the other Methuselahs, for they were among the longest-lived creatures ever to evolve. Although the ammonites became outdated, were rendered largely obsolete by the evolution of shell-breaking carnivores, their history was one of such uncommon and clever adaptation that they *should* have survived, somewhere, at some great depth. The nautilus did. It is my prejudice that the ammonites would have, save for a catastrophe that changed the rules 66 million years ago. In their long history they survived everything the earth threw at them. Perhaps it was something from outer space, not the earth, that finally brought them down.

6

Timeless Design
The Horseshoe Crabs

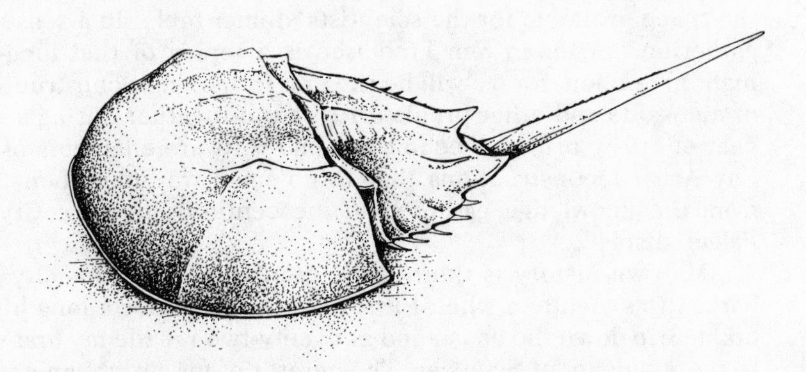

A Gathering of Scientists

San Francisco must be the most beautiful city in the world, especially as you drive in on a perfect late-April afternoon, moving freely across the Golden Gate Bridge against rush-hour traffic as a strong wind whips the bay to whitecaps. I'm eagerly anticipating the evening—not a black-tie affair (scientists rarely have them), but close: the California Academy of Sciences is holding its annual Fellows Night, a yearly get-together for the scientists of the Golden State, and this year the evening's festivities promise to be extraordinary. We're being given the first peek at the academy's newly constructed Hall of Evolution, followed by dinner at the Steinhart Aquarium.

I reach Golden Gate Park in a dead heat with the setting sun. The Rhododendron Glen is in full bloom, and as I drive past the conservatory I'm reminded of the famed Crystal Palace in London, where a century ago another gathering of scientists witnessed an extraordinary scene: the great English anatomist Richard Owen had conceived and executed life-sized reconstructions of the then newly discovered dinosaurs. Giant statues of these creatures were scattered about the grounds surrounding the Crystal Palace, a large, glass-paned conservatory that looked much like the conservatory that now stands in San Francisco's Golden Gate Park. Owen and nineteen other guests even dined within one of the reconstructed dinosaurs, the first model of an iguanadon, which roamed southern England over 100 million years ago. The reconstruction looked much like a giant frog, for the bones had only recently been discovered and were far from complete; only after much subsequent work and new discovery would the scientists of the time find that this giant herbivorous dinosaur walked erect on two hind legs. The squat shape of the reproduction had its advantages, however, for it increased the space available for the scientists' dinner table. In a sense the gathering tonight in San Francisco is a replay of that long-ago night in London, for we will be the first to see the reconstructions of dinosaurs and other prehistoric life put together during a five-year effort by brilliant technicians and software engineers of the Bay Area, reconstructions that in no small way have benefited from the knowledge gathered in the century since the Crystal Palace display.

My own history is thoroughly intertwined with this city and park. I was eighteen when I first visited here, after a long hitch-hiking trip down the coast, and at twenty-two I made my first visit to the Academy of Sciences. The curator of fossils, a man named Peter Rodda, admitted me to look at the extraordinary collection of ammonites and fossil nautiloids kept there. Now I'm back, two decades and much research later, on a honeyed evening.

I mingle among the grayer heads. The assemblage makes me feel very young, but I see great friends scattered among the Nobel laureates and National Academy members: the paleontologists Jere Lipps, Bill Clemmons, Leo Laporte, and, best of all, my old friend Peter Rodda, the mastermind of the project, more bowed but still gracious. We gather in the planetarium to hear the history of the hall. It was conceived as a statement, an affirmation of the principle of evolution. Millions of dollars have been spent and five years have elapsed since its inception to complete this bold statement of our acceptance of Darwin's theory. There is no sop to creationism here; only science is on display.

When we finally view the hall itself I am stunned; many ancient worlds have been brought back to life by ingenious technology. I walk through the Carboniferous coal swamp to see the creatures of that world, the first reptiles and ancient amphibians, the giant arthropods all in motion and seemingly alive. The effect is heightened by the living plants around the exhibits, species of those long-ago ferns and cycads, horsetails and pines that lived then and live now, Methuselahs on display. I proceed into the

Mesozoic wing, where fearsome dinosaurs move among the pine forests and tiny mammals hide in the wings, gnashing their teeth in defiance, awaiting their turn. I pass into a large alcove dedicated to the Mesozoic marine world and find old friends there: fossil ammonites that I've collected from Mill Creek in Northern California. The greatest treasures I've extracted from the fossil record are no longer hidden in the museum drawers where I left them many years ago but are now on permanent loan to the academy, to stir the wonder of visitors. In a giant diorama they have been brought back to life, drawn into being once again by the genius of Peter Rodda, the creatures of my dreams now reconstructed with tentacles and colored shells and the power of being, frozen in mid-water with the other denizens of their world, the flat clams and shell-breaking fish, the marine lizards and snails. The effect is breathtaking, and I have to congratulate the small crew that put this giant diorama together, for they have snatched back from time the sea that once covered Mill Creek. I'm amazed that the workers who have constructed this exhibit await my judgment on their vision. They seem relieved beyond measure to see my joy. An entire world I have visited many times has been brought back to life.

At the end of the evening, alone now, I stroll one last time through the gallery, and notice an aquarium I missed on my previous visits. Like any other museum, this one has far too many exhibits to see on one visit. I walk up to the brightly lit tank and see the most extraordinary creatures. They seemingly defy gravity, swimming languidly on their backs. Some do cartwheels and underwater acrobatics, all in slow motion. They're small for their species, juveniles still long from maturity, with many molts still to go before they reach their full size. The tank is filled with horseshoe crabs, common and familiar to beachcombers of the east coast of North America, but unknown to us Westerners. They look like ungainly tanks but they move without effort in the water, either crawling on the bottom or swimming above it; I marvel at their swimming prowess. They have existed on this earth a very long time, since early in the Paleozoic. It's fitting that they live here now, in this Hall of Evolution, some of the most long-lived survivors on this planet.

Slaves to the Moon

It's hard to imagine our world without its moon. Countless songs, poems, and other flotsam of our culture would never have been, of course. But in a biological sense, the moon may have very little to do with our human ancestry and ecology. If we were less passionate, our species would in all probability be not much different in a moonless world. Such is not the case for countless other species, though. Some species are slaves to the moon, and probably have been for very long stretches of earth history.

Limulus polyphemus, the common horseshoe crab, follows the incoming tide onto the sandy beaches and mud flats of the East Coast, feeding on worms and bivalved clams. These crabs forage under the cover of the incoming tide and retreat with the ebb, for they are fully aquatic creatures and have no interest in life on land. During one time of year, however, the horseshoe crabs move into the shallows with intentions beyond just feeding. When the moon and sun combine their power over the ocean to produce the annual spring tides, when the ocean makes its annual bid to engulf the land with the highest tides of the year, the horseshoe crabs come as well, moving upward and inward over the shallow underwater shelf, thousands upon thousands of them driven into the intertidal region, until they've gone as far as the tides can reach. There the females scoop out nests in the sand and lay their eggs. The eggs are immediately fertilized by an attending male, which in most cases has ridden onto the beak atop the female. She is larger than he is. Then, usually under the cover of night, the crabs follow the now-retreating tide downward, back to the sea. The eggs stay buried in their deep resting places, developing slowly, and will not hatch until the tide once again rises upward to engulf the nests some months later. The newly hatched larvae emerge from their nests and grow slowly for a year in the intertidal zone, freely foraging when the tide is in, burrowing into the sand for protection from desiccation and predators when it ebbs. Gradually they grow, periodically molting their hard exoskeleton, and as they grow they move outward from their intertidal home to explore ever-greater depths in the sea. Eventually a time comes when they leave the sand flats of the beaches, bays, and estuaries to stay offshore in the subtidal regions, now living in a world never exposed by even the lowest tides. Finally, after five to ten years, they reach their full size. And with attainment of maturity they soon feel the pull of the moon. Like their parents before them, they wait for the rising spring tide to renew the cycle, unbroken for tens and perhaps hundreds of millions of years.

Horseshoe crabs are amazingly tough. They can withstand days of drying if they are cut off from the sea; their system can function in water that varies widely in salinity, from normal sea water to water that is nearly fresh. They can withstand great swings of temperature and can live on virtually any type of prey. They are supreme generalists, and seemingly could live on virtually any shore of the earth. Their hardiness is legendary. So why isn't the earth awash in horseshoe crabs? Why can't they be found on every shore? Why are they found only in shallow water? And with their ability to live for extended periods of time out of water, why didn't they take the final step and, like their close relatives the scorpions and spiders, colonize the land?

The Man Who Studied Horseshoe Crabs

I grew up in the 1950s, fed on a constant stream of movies in which humans confronted extraterrestrial creatures. Even then I wanted to be a scientist, and I chafed at the image of scientists offered by the Hollywood of that time. There seemed to be two contrasting stereotypes: (1) wise old man with beautiful daughter who helps slay monster, but invariable loses daughter to macho military hero (the wily old man in *Them* comes to mind), and (2) well-meaning but naive and ultimately foolish man who wants to communicate with monster and gets toasted for his trouble. The rather slimy individual in the original version of *The Thing* is the perfect example: he finally suffers the inevitable squashing at the hands of the eight-foot vegetable from outer space and the military man gets the girl. My motives at this early time were quite clear: I wanted to be the scientist *and* get the girl. It was thus disillusioning when I finally went to my first large scientific meeting and saw for the first time the actual physical embodiments of the giants in my chosen field, geology. Without ever actually inventing a face, I had devised a mental picture of the men and women who had written the important papers that had become the building blocks of my scientific knowledge. Unconsciously I had given them heroic features.

Such thoughts were crossing my mind in Rochester, New York, on a cold, snowy night in 1976. Two months earlier I had returned from my first and, as it turned out, longest trip to the South Seas to study the chambered nautilus. This four-month voyage had produced notebooks full of numbers and observations. I returned from the tropics to complete my last year of graduate school.

I was intrigued by the findings I had made during my voyage to New Caledonia, and wanted to present them to a group of paleontologists outside of my own university. I therefore made a nervy phone call to the most famous paleontologist in the United States, a man named David Raup. Raup was a particular intellectual hero of mine. Although I had never met him, I felt I knew him well from immersion in his published work. He was largely responsible for a revolution that in the 1960s transformed paleontology from a field concerned almost entirely with the recognition and definition of new species of fossil plants and animals into a twentieth-century science concerned with rigorous hypothesis testing and mathematical proof. As one wag from *Nature* put it, Raup helped lead paleontology back to the high table.

So I rather bravely dialed, was connected to the great man, and proceeded to invite myself to his university for a seminar. To my surprise, Raup readily agreed, and two weeks later I was crossing the frigid expanse of upstate New York in a Greyhound bus bound for Rochester.

At the time of my visit, Dave Raup had managed to persuade his department to hire two young paleontologists for the geology faculty at Rochester. This was no easy feat, for most geology faculties have room for no more than one representative of each of the myriad subdisciplines of geology. At the time of my visit Raup had assembled a large and vibrant group of graduate students as well as two newly minted Ph.D.'s, both recently arrived from Harvard, named Jack Sepkoski and Dan Fisher. I remember being very surprised when I met them for the first time. I had heard of both of them and read their works. It hadn't occurred to me that these two scientists, already well known, were still in their mid-20s, about my own age. And both looked like normal people.

Since then most of the graduate students at Rochester have become well-known and respected paleontologists, and the two young professors have moved on to other universities and are now establishment figures. Jack Sepkoski is one of the premier paleontologist/statisticians of my field, making important contributions in many areas of paleontology. But the man I remember most vividly is Dan Fisher.

I gave my seminar that night, detailing my findings on the buoyancy system of the nautilus and its relation to the workings of the long-dead ammonites. Afterward we adjourned to a large common room for questions and discussion. Someone had negotiated several large pizzas, and I was in the postseminar haze of wanting to eat and not wanting to talk until I had decompressed. I was a novice at the seminar game, and nervous before this large and sophisticated audience. I remember falling into a large couch and reaching for a slice of pizza as the first questions began. I had no problems until I had to field a question from Dan Fisher. There was a sweet, luminous innocence to his face, and it was clear that the only motive behind his questioning was curiosity. I've been among countless academics since then and have witnessed intellectual brutality often. Some questioners adopt an attack style in efforts to belittle the victim. Such people, I believe, can mask their own insecurity only by humbling others. Fisher's questions were not at all of that sort, yet they shook me up, for he had the sharpest intellect I had ever encountered. Question after question forced me to think deeply. He examined my major points and turned them over to look at them in new ways; and when he found errors in my reasoning, he gently exposed them. I finished the night in despair. One of the other graduate students clapped me on the back. "Don't worry," he said, "he does it to nearly everyone. Dan's just a lot smarter than most people."

A year later, at a conference of geologists from all over the country, I saw Dan Fisher again. I finally got a chance to ask him about his own work. He smiled shyly and told me that for many years he had studied just one creature. And then, for the next several hours, he held me captive by his stories about the life and history of horseshoe crabs.

The Evolution of the Horseshoe Crabs

The origin of the horseshoe crabs is obscure. Three small, fragmentary fossils from early Cambrian rocks, sediments deposited over 550 million years ago, may represent their earliest known ancestors. The creatures that yielded these tiny clues to the origin of this ancient group probably looked much like the most common animals of the time, the trilobites, arthropods characteristic of the Cambrian Period. But the fossil material is so rare and fragmentary that we can only guess whether these tantalizing fossil glimpses really mark the first appearance of the horseshoe crabs. We next glimpse creatures considered to be ancestral to the living horseshoe crabs in Silurian-aged strata, sediments of somewhat over 400 million years ago. Here the evidence is better: the fossilized skeletons of these creatures are relatively complete. They look like horseshoe crabs in the head region, but have more body segments and a different tail region; they thus seem to be good transition forms, bridging the gap between the earliest ancestors of the horseshoe crabs and the forms that are characteristic of the living species. A variety of forms evolved in the Silurian Period, lived a short time, and then went extinct. Those that survived or gave rise to new species had a common trait: they grew steadily larger and the segments of the body gradually fused into a solid plate. By the Devonian Period, about 400 million years ago, the group had a structure clearly recognizable as characteristic of the horseshoe crabs, and by the Carboniferous Period, about 300 million years ago, all of the pieces were in place: the horseshoe crabs had arrived at the stable, successful shape still found among the living species of today.

The Carboniferous Period, between about 360 to 300 million years ago, appears to have been the heyday of the horseshoe crabs. Many species evolved and flourished in a variety of habitats. After the Carboniferous Period, however, the tribe of horseshoe crabs went into decline and the number of species dwindled. Such a loss of diversity can often be the first sign that a group is heading toward extinction, that more successful creatures or adverse environmental conditions are squeezing these creatures out of the living world. But the trend followed by the horseshoe crabs was curious: as the rate at which new species formed declined, the rate at which the living species became extinct declined too. The result is that once a horseshoe crab species evolved, it became very long-lived; only rarely and after many millions of years did a species go extinct. The fossil record of these animals is never plentiful, and there seem never to have been more than a handful of species on earth at any one time. Today only four recognized species are still living. But this number has been relatively constant for hundreds of millions of years, and for most of that time the horseshoe crabs have been living lives very similar to the ones they lead today. They continue to live in the shallow, often brackish water of bays and estuaries, and spend most of their lives on the shallow subtidal and intertidal sand and mud flats of the seashore. Although the group began its history as creatures of the bottom of the open sea and in all probability lived in water of normal salinity, they appear to have adapted quite rapidly to the harsh life of the seashore, where the salinity, temperature, and oxygen levels fluctuate. The horseshoe crabs have had to evolve elaborate physiological mechanisms to enable them to withstand dry periods and wide swings in salinity. Doubtless these adaptations did not come immediately or all at once, but they probably did come over the course of the Carboniferous Period, the time of their greatest diversity. When that period ended, most of the species died out. But those that survived were granted the gift of great longevity—the closest thing to immortality offered by the biosphere of our planet.

During the Carboniferous Period the horseshoe crabs dared an audacious experiment. Living in shallow water, as many of them did, and probably being able to withstand at least short periods out of water, at least one species made the attempt to colonize the land. It did so in the huge swamps of North America, an environment that has yielded one of the greatest natural resources on earth: the mid-continental coal deposits.

The fossilized remains of the soft parts of extinct animals are as rare as their skeletons are common. The three most famous places where they may be found—the Cambrian-aged Burgess shale of British Columbia, the Carboniferous-aged Mazon Creek of Illinois, and the Jurassic-aged Solnhofen limestone of

Germany—have yielded thousands of fossilized creatures found nowhere else on earth. They are our best windows into the past, the closest we have come to time travel. In each of these places the bottom mud appears to have contained very little dissolved oxygen: Consequently, when animals died and sank into the mud, they came to rest in graves where scavengers could not tear them apart and bacteria could not decompose them; essentially, they became mummified. And as time went by, the soft parts of their tissues became impregnated with minerals that differed slightly in chemical compositions from those in the surrounding sediment. When these fossils are exhumed from the rock, they show the traces of their soft parts.

Though the Burgess shale, Mazon Creek, and the Solnhofen limestone differ in the age and structure of their rocks, all three contain the fossils of either horseshoe crabs or, in the case of the Burgess shale, arthropods very much like them.

The Mazon Creek Fauna

About 300 million years ago, in what is now Illinois, a wide, flat plain extended into a warm, shallow sea. This plain, a large delta, was built by the slow accumulation of sediment brought by the several large rivers and their tributaries. The delta was covered with a riot of vegetation, mostly plants that would seem very strange to us: club mosses and seed ferns, giant horsetails and archaic pines. The animals would seem no less strange: giant centipedes and dragonflies, and myriad smaller insects chased by hungry amphibians. The trees grew very rapidly in the rich soil, becoming giants quickly and then toppling over in the soft muddy plain to add to the huge piles of vegetation rotting in the swamps at the edge of the delta. The blood-red water of the nearby sea was murky with the muddy discharge of the sediment-laden rivers. Like the swamps, the shallows of the sea were rich with the life that fed on the nutrients that flowed out of the rivers. Untold numbers of jellyfish filled the seas, as did a variety of worms, mollusks, and fish.

Many such worlds existed in the past, but few can be so well documented as the Carboniferous Period world of Illinois. The delta and the nearby sea have been preserved in stone along meandering creeks and quarries. Scattered within these shales and sandstones are extraordinarily hard stones aptly called ironstones. Sometimes more than a foot in diameter but usually smaller, sometimes round but more often oblong, these curious objects will split apart with a satisfying crack if they are hit hard enough with a sledgehammer. And if they are hit well, they will usually split cleanly along the long axis, and more often than not will yield paleontological treasure. For more than a century, geologists, professional and amateur alike, have been splitting open the ironstones of the shales along Mazon Creek and elsewhere in Illinois to find the extraordinary faunas within them. More than 500 species of animals and plants found in this way have now been identified. It is one of the most extraordinary and diverse fossil faunas known anyplace in the world.

The Mazon Creek fauna, as the assemblage of fossils from this area is called, has been preserved because of a series of geological coincidences. As a result of the chemistry and depositional characteristics of the sediments of the delta and the nearby sea, the animals and plants of this region were quickly buried and encased in rock after their death. Most of the creatures interred in the soft sediment were so rapidly entombed that the normal complement of scavengers and decomposers had little chance to break down the tissues of the dead. Structures of phosphate and cuticular material became particularly well preserved, while, paradoxically, calcareous shells dissolved away in the acidic sediment, in a process just the opposite of the normal preservation cycle. The result is that creatures familiar to us in our world but vanishingly rare in the fossil record—worms, insects, jellyfish, and spiders—came to be preserved in abundance in the ironstone concretions of the Mazon Creek strata.

One of the most exquisite fossils to be found in the Mazon Creek assemblage is a small horseshoe crab covered with spines. Two of the three horseshoe crab species known from the various Mazon Creek fossil collections are extremely rare; in fact, they are known only from deposits formed in the shallow sea of this area. The third species is far more common, and is found with creatures that usually live in a very different environment. This tiny creature has been given the odd name of *Euproops*. It became the object of Dan Fisher's attention, and under his unrelenting gaze finally yielded an interesting secret. This small horseshoe crab, unlike its ancestors and descendants, appears to have been attempting to colonize the land. The fact that the world is not awash with terrestrial horseshoe crabs attests to its ultimate failure in this endeavor. All the same, its attempt is an interesting story in itself.

Long ago, in the Paleozoic Era, the arthropods invaded the land on four occasions. First the scorpions and their relatives the spiders established a beachhead. The scorpions seem to have been among the first larger creatures to venture out onto the early Paleozoic landscape, for their legs and shape, though evolved for life in the sea, were superbly functional on land as well. A second group was the insects, now among the most successful of all land-dwelling creatures. The third group of invaders consisted of the isopods, known to us as the pill bugs or potato bugs, actually small crustaceans. The fourth attempt, independent of the others, seems to have been made by the horseshoe crabs.

Dan Fisher arrived at this conclusion on the basis of both the morphology and the faunal associates of the tiny Mazon Creek horseshoe crab species *Euproops danae*. Fisher found fossils of this species, unlike those of the other two species, to be associated with creatures known from land and from fresh water, not seawater—insects, spiders, land plants. He even found some of these horseshoe crabs nestled in the fossilized trunk of a large tree. This is the sort of habitat favored today by the pill bugs, which seek dark, moist places out of direct sunlight. But the most telling evidence, to Fisher's mind, was the presence of spines on the *Euproops* fossil. Spines tend to be a means of defense, an armor that discourages predators. In the case of *Euproops*, however, the spines on the carapace may have served as camouflage. One of the most common plants in the swamps and ponds of the ancient Carboniferous Period deltas were the lycopods, such as

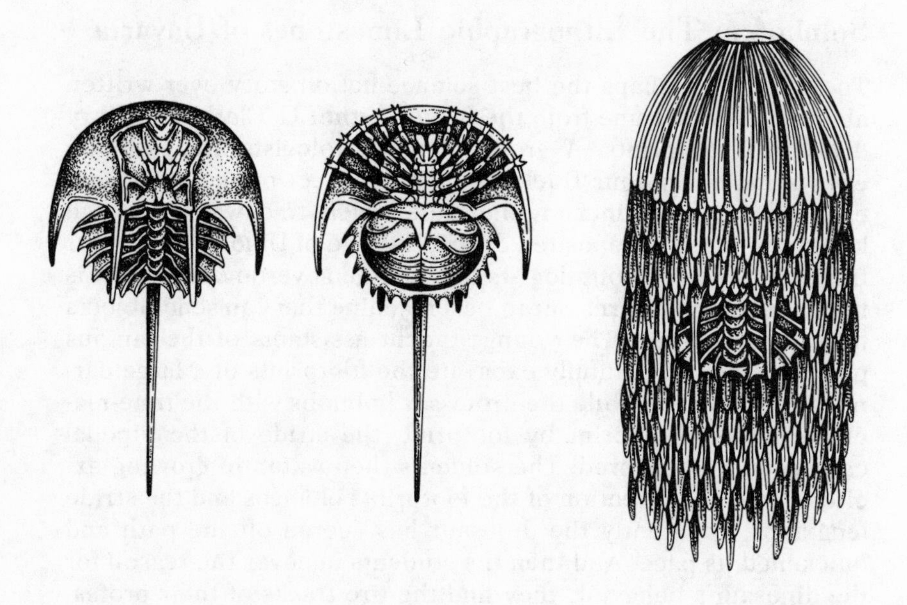

The horseshoe crab Euproops *from Mazon Creek. The left and middle figures show the crab as it is thought to have looked in life; at the right a crab clings to a Lepidodendron twig in a coal swamp, camouflaged by its spines. (Adapted from Dan Fisher, Evidence for subaerial activity of* Euproops danae, *in* Mason Creek Fossils, *ed. Matthew Nitecki. New York: Academic Press, 1979, pp. 379–447, by permission.)*

ground pines. These trees had scaly, spiny trunks. Fisher demonstrated that the spines of these horseshoe crabs closely matched those of the lycopod trees in size and shape. A horseshoe crab that climbed onto one of these trees would be essentially invisible to its predators and perhaps to its prey as well. The tiny crab could lie in wait, well hidden, to pounce upon unsuspecting prey. With legs prefectly shaped to cling to tree branches, this species, alone among the horseshoe crabs, appeared to have made the transition from water to land.

There are no land-living horseshoe crabs in our world. Nor is there any evidence that the horseshoe crabs actually succeeded in becoming complete land dwellers in the Paleozoic. Why did they fail? Perhaps they arrived too late. If the horseshoe crabs had made their clumsy way onto shore in the early Paleozoic, they would have encountered an empty environment and perhaps would have flourished. In the Carboniferous Period, however, they wandered onto an environment already filled with predators. By escaping the wicked teeth of the Carboniferous Period marine fish, the horseshoe crabs climbed out of the sea into the waiting jaws of the voracious amphibians.

Solnhofen: The Lithographic Limestones of Bavaria

To my mind, perhaps the best science-fiction story ever written about dinosaurs came from the pen of Arthur C. Clark. The story deals with footprints. A group of paleontologists are patiently excavating a dinosaur trackway, a huge sheet of sediment covered by the footprints of dinosaurs. Such trackways are well known from strata deposited during the Age of Dinosaurs. Not far from where the paleontologists toil, a large government project is humming away; it turns out to be a machine that can send objects back through time. The young student assistants of the famous professor slowly, painfully excavate the footprints of a large carnivorous dinosaur while the professor hobnobs with the time-machine people. Footprint by footprint, the stride of the bipedal carnosaur is uncovered. The students then watch in growing excitement as the direction of the footprints changes and the stride lengthens; apparently the dinosaur has veered off the path and quickened its pace. And then the students uncover the reason for the dinosaur's behavior: they find the tire tracks of their professor's jeep, now preserved in Jurassic-aged sediment amid the dinosaur footprints. The jeep, with professor aboard, was presumably sent back through time, and unfortunately caught the dinosaur's eye. The story ends just as the students are about to excavate the remains left by the presumably bloody encounter between the pursuing dinosaur and the fleeing jeep.

I first read this story (and everything else Clark wrote) as a teenager; I stumbled across it again this year, and once again enjoyed it immensely. But on the more recent reading the story brought to mind one of the most famous of all known fossils: the trackway of a horseshoe crab preserved in one of the most exquisite of all lithic media, the limestones of the Solnhofen quarries in southern Germany. In Clark's story we never see the fossil of the body found at the end of the trackway; at Solnhofen, however, we get to see the final act of the tragedy.

The stones speak of more than the skeletons they preserve; they tell us about ancient behavior as well. Living animals usually engage in a variety of activities and movements that leave traces in soft sediment: trackways, burrows, and feeding traces are all commonly preserved in sedimentary rock. Trace fossils of this sort can be exceedingly useful to geologists in their attempts to reconstruct sedimentary environments. Just as the assortment of animals changes as one descends deeper into the sea, for instance, the assemblage of trace fossils preserved in sediment tends to vary with the depth of the water. The trace fossil material found in ancient rock can often be a powerful clue to the depth of the sediments that were deposited. Besides, trace fossils often provide the only record of creatures that had no skeletal hard parts. The bodies of worms, for example, are practically absent from the fossil record because they had no hard skeletons, but the trace fossils they produced are very common.

The problem is that it's usually impossible to determine precisely which species of animals made which trace. A marine worm moving through mud usually leaves a record of its movement; so does a crab walking across a fine muddy bottom. The trace fossil left by the worm will often be preserved as a tubelike structure meandering through the strata, and the trace fossil of the crab's movement will look like a series of tiny grooves across the top of the sediment. In either case the trace fossil left behind leaves indisputable evidence that some worm or crab was present—but which worm or crab? There are many thousands of species of both, and the traces they leave behind usually look the same. The actual animal that produced the trace can be identified only in the most extraordinary circumstance. Just such an extraordinary trace was discovered in the Solnhofen limestone, a fossil deposit that has yielded some of the world's most priceless fossils, including the only known specimens of Archaeopteryx, the first known bird fossil.

Many years ago the workers at the limestone quarry found a long trackway that recorded the passage of some crustacean across the face of the 150-million-year-old sediment. The limestone here is covered with such trackways, for animal life was common in the shallow lagoon in which these sediments were deposited. But at the end of this trackway lay the remains of the animal that produced it, a horseshoe crab, beautifully preserved. Near death, the creature had walked across the seabed, and then stopped and died. Sediment settled over both the track and the animal. The fossil is priceless.

Dan Fisher examined this fossil and the remains of other horseshoe crabs preserved in this quarry. To a less critical eye, the horseshoe crabs of that long-ago time look virtually identical to the present-day species. But Fisher found slight differences in the carapaces of the Jurassic and the modern species, and investigated how these differences would affect the animals' swimming. He found that the slightly flatter shape of the Jurassic species would affect the orientation of the crab in water. Modern horseshoe crabs usually swim upside down, with the tail lower

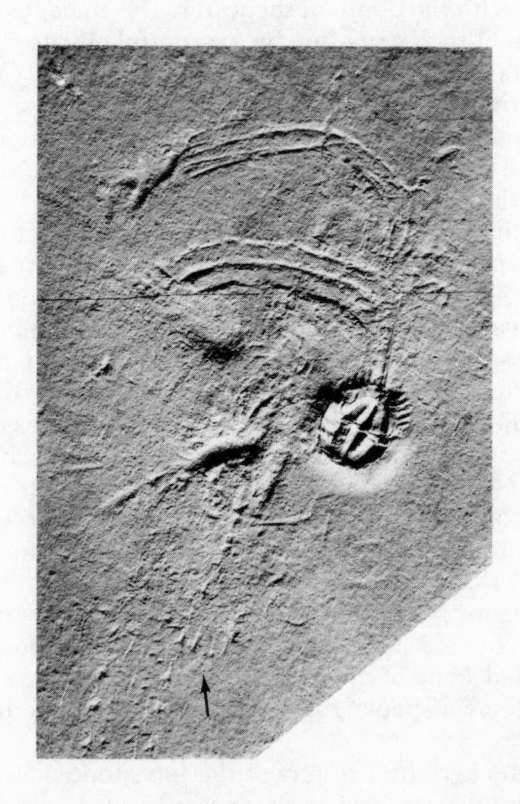

A horseshoe crab and its tracks preserved in limestone at Solnhofen, in Bavaria. (H. Leich, Aufschluss *1:5–7, 1965.)*

than the head, using their legs to "row" through the water. The slightly flatter Jurassic species probably swam in the same fashion, but their flatter shape would have permitted them to swim horizontally. The trackways of the Jurassic species have also yielded valuable clues to the life of these ancient horseshoe crabs. By analyzing the length and shape of each of the footprints, we can also understand the way these crustaceans walked. We know now that at least since the Jurassic Period, the middle of the Age of Dinosaurs, the appearance and probably the way of life of horseshoe crabs have changed hardly at all. They are a creature in stasis. Empires of animals have come and gone, from giant dinosaurs to Ice Age mammals; predators in the sea have become quicker, smarter, and very adept at killing even the most heavily armored creatures; and perhaps the skies themselves have fallen, as giant meteors have smashed into the earth and obliterated much life. But the horseshoe crabs—never more than a small handful of species at any one time—have gone on. How is success determined in the history of life? By the number of species produced? Or by the longevity of the group? Who could call the horseshoe crabs unsuccessful?

Woods Hole

Two months have passed since my meeting in San Francisco. I am once again at a scientific assembly, but of a different kind, in a vastly different place. In contrast to the exuberant energy of San Francisco, I am now surrounded by the understatement of New England. I am attending a meeting of malacologists—scientists who study mollusks—at Woods Hole, Massachusetts. Five days of scientific interchange have just ended. For many hours we have listened to paper after paper about the biology of squids, gastropods, and clams. It's nearly time for me to go; my bags are piled up around me as I await the bus back to Boston. I stand on the pier and look over the side at the shallow sandy bottom below, my mind filled with the events of the past week, of people met and science communicated. And as I look at the sea below me, part of the sandy bottom is transformed into a large horseshoe crab. I have to laugh. Last night a large contingent of the assembled scientists scrambled into vans to go in search of horseshoe crabs. On a beach near the famous Woods Hole Oceanographic Institution, an Earth Watch program was studying the mating of the horseshoe crabs. Several nights ago, under a full moon that coincided with the highest tides of the summer, the volunteers reportedly had counted nearly a thousand crabs climbing onto the beaches to breed and lay eggs. Night after night in the moonlight the crabs returned. So we malacologists decided to see this wonder of nature for ourselves. But the night we chose was stormy. We blundered among the million-dollar homes overlooking the seas hoping to find a beach sheltered enough to be free of pounding surf. Nary a crab did we see. Some wag pointed out that we were malacologists, after all, and what did we expect? All the same, we were disappointed.

Now I gaze down at the crab moving slowly below me toward the beach, oblivious of my presence. We are like ships passing in the night.

The horseshoe crabs have survived a very long time, for readily understandable reasons. Essentially they are all armored head, difficult for any predator to assail. They live along the seashore, and, like the brachiopod *Lingula*, have adapted to this harsh environment of changing salinity, temperature, and food supplies. In sum, they seem to be extraordinarily hardy creatures. The paleontologist Niles Eldredge of the American Museum of Natural History has even pointed out that the horseshoe crabs are extremely tolerant of chemical pollution. In the estuaries and bays of the eastern United States, horseshoe crabs are often the last living creatures to be found amid the poisons and effluents generated by our industrial zeal. Perhaps this trait will become the most important factor in determining which species will continue to be living fossils.

Only a few horseshoe crab species still exist. Yet they have had a great ride. They owe their survival to the lucky chance of evolving a form fit for the ages, a form that worked in the long-ago Paleozoic and still lets them live today, in a world filled with carnivores and competitors unimaginable to the Cambrian world.

7

The First Spring
Plants Invade the Land

The Salinas Formation

Standing on the cobbled shore, I look out to the horizon, across the sea before me, for a sea is what this huge expanse of water looks like. We flew in from Columbus this morning, our tiny Cessna crossing the flatness of northern Ohio into Michigan, skirting the more ominous-looking cloudbanks on a northern course, finally to descend along the edge of this giant lake. I had seen the Great Lakes before either from a jetliner at 30,000 feet or from a car, but never from this altitude. They seem like oceans, an endless expanse of wavetops. We landed on a grassy strip and made our way to the lakeshore, a cheerful crew. Most of my companions have come on this mid-May excursion as a lark, an excuse to escape Ohio, a fine way to spend Derby Day. But as usual I have ulterior motives: once more my real destination is a rock formation. A cool wind blows across the huge lake, sending small waves lapping at my feet as I search the pebbles around me for telltale signs. These cobbles are the eroded remnants of the Silurian-aged Salinas Formation, a unit of sedimentary strata originally deposited more than 400 million years ago. North America was moving eastward then, and the Appalachian Mountains were beginning to rise along the eastern seaboard. As the mountains began to form to the east, the more stable interior of the continent was warped ever so gently by the far-off compression into vast crustal arches and depressions. A giant basin, hundreds of miles across, formed in the area someday to be called Michigan, connected to the Atlantic coastline by a wide shallow sea. Flourishing coral reefs formed along the edges of this basin, while in the middle, at greater depth, the muddy sediments were home to a rich assortment of brachiopods and other Paleozoic life forms. This idyllic place existed for tens of millions of years. But as it has done so often over the millennia, the environment finally changed in a way that killed off most of the reefs and marine life. All over the world the level of the sea began to fall; eventually, the Michigan basin was isolated from the open ocean. The hot sun began to work on this inland sea, evaporating its water. As the sea sank, the water became increasingly salty. Huge salt deposits were left behind as the water continued to drop away. Here and there around the margins of the basin, swampy areas formed. Here an odd assortment of life evolved. Species descended from fully marine creatures became adapted for life in brackish water, in the lagoons and deltas and river mouths that made up this area so long ago. It is the fossilized remains of this fauna that I seek, on these gray cobbles washed by Lake Michigan.

I kick over innumerable pebbles in the May sunshine, an odd beachcomber, looking for signs of late-Silurian life. I find few fossils, and for my companions, now all pleasantly buzzed on the local concord grape wine, they hold little interest: scraps of arthropod exoskeleton and tiny flecks of bone. But I treasure them all the same, and finally I find a trophy they consider worth a look: on a large slab I find a claw, many inches across, a cruel pincer studded with sharp needles. This ages-old fossil is clearly a killing organ, a weapon designed to catch and rend flesh. It's the claw of a demon called a eurypterid, a fortunately long-extinct creature whose closest living relative is the scorpion. Most eurypterids were small creatures, at most several inches long, spiny arachnids living in the lagoons and estuaries of the Silurian seas. But some eurypterids were giants. The largest fossils yet discovered were over six feet long, huge beasts with fearsome claws, creatures evolved and adapted to feed on one succulent type of prey: us. They are the creatures of our nightmares, stalkers of our earliest vertebrate ancestors, for it was in these Silurian waterways that vertebrate evolution began, the first stirrings of the ancestral fish design, tiny bottom feeders, necessarily furtive among these gigantic killing machines. The eurypterids changed the course of vertebrate evolution, requiring our earliest ancestors to adopt heavy, bony body armor. Perhaps, deep within our genetic code, their memory lurks still. We flee from the scorpion or the spider in the recesses of a dark closet, a result, surely, of our species' interactions with these often poisonous creatures. But our fear of them may come from wells far deeper than the mammalian sections of our brain. The monstrous eurypterids once stalked us in untold numbers, and they may have been a major reason that some early fish finally crawled from the sea.

An aching back makes me straighten up, and I retire to a comfortable log at the edge of the beach to wrap and label these precious slabs. I have to look at my treasures one more time before they find their way into field wrappings, and in this process, amid arthropodan remains on one of the collected rocks, I spy an unknown trace of previous life, a tiny layer of organic material. I look at it, puzzled at first, until the fossil resolves itself into plant material. Through my hand lens I try to make out internal structure, but can see only that it's from some sort of vascular plant, a plant with a system of channels to transport water through its body—a tiny land plant. I don't know it at the time, but this tiny smear is perhaps even more precious than the traces of animal life I've collected. The tiny plant fragment has come from land, Silurian-aged land. It's a trace of a truly unknown soldier, remains from the first invasion of land by the plants, a pioneer of the invasion that would completely transform the surface of the earth. While our ancestors dodged and fell prey to the eurypterids, the greatest conquest of all was taking place: the conquest of the land by the plants, planet-shapers that ultimately made the soil and produced the oxygen and greened the land, creatures that made possible the eventual colonization of the earth's surface by animals.

The First Wave

The timing of the first wave of arrivals in the conquest of the land has been the subject of lively debate for many decades. Lost in the discussion of when the first plant immigrants came ashore is perhaps a more perplexing question: Why? Why colonize the land at all? For surely it would be hard to imagine a more inhospitable place than the land of the earliest Paleozoic Era.

The environment colonized by the first land plants must have been harsh indeed. The land probably resembled the most sterile portions of the great deserts of today, totally devoid of plant life, a place without soil, for it's plants that create soil; it was a place of harsh winds, shifting sediment, and steadily marching sand dunes. There would have been much less cloud cover and precipitation, for the plants of today have a large effect on the recirculation of water vapor back into the atmosphere. Land that lacks plant cover has scant cloud cover and little rain. When rain did fall on this ancient land, it would often run off quickly, picking up and carrying unconsolidated sediments along in flash floods, or sheet-wash. With no roots or plant cover to stabilize the upper layers of sediment, there would have been constant movement of rock and sediment on the earth's surface. It was onto this harsh habitat that the first plant pioneers emerged from the sea. Why did they come? To escape the fierce competition of the shallows of their origin, or to dodge the increasing numbers of plant-eating herbivores evolving in the shallow seas? Or simply because a huge new environment offered new resources and living space? We cannot know.

We have few fossil remains from the first terrestrial colonists, and thus must resort to speculation about these first invaders. Perhaps the earliest arrivals of all were hardy fungi, living as smears on the rock surfaces, or simple algae living in damper areas. Or perhaps the earliest of all were single-celled plants, living in wetter areas. The first true multicellular land plants, forms with a vascular system to transport water through their bodies, are not known from terrestrial sediments until sometime in the Silurian Period. This period started about 440 million years ago and ended about 400 million years ago. Paleobotanists have concluded that these earliest vascular plants must have originated or descended from green algae. To make the jump from a marine lifestyle to a life on land, the earliest land plants had to evolve numerous new structures—changes in form not only to protect themselves against desiccation but also to gain access to water and nutrients and to reproduce and disperse successfully.

Perhaps the greatest danger faced by the early land plants was desiccation. The harsh land environment with its incessant winds and sporadic rainfall necessitated some system to keep the plant from drying out. To maintain their moisture levels the early plants evolved a thickened outer cell layer, called a cuticle. This layer did stop water loss, but it also restricted the exchange of gas between the plant's interior tissues and the atmosphere. To solve this problem, the plants opened small holes in the cuticle.

With the evolution of a cuticle, it was but a small step to evolve an upright growth form. Until now plants had sprawled on the soil or on rocks; the cuticular system stiffened the plants and allowed them to stand more nearly erect. As the first wave of early immigrants spread over the land surface and competed for sunlight and a toehold for their reproductive spores, they grew taller and taller. Greater height led to another problem, however: when plants no longer hugged the ground, they needed some way to transport water and nutrients throughout their bodies. This problem was solved by the evolution of a vascular transport system.

The most vexing problem of all was a method of reproducing and dispersing. The multicellular, sexually reproducing plants of the sea, such as the green algae, reproduce by means of zoospores—cells with long, tapering appendages by which they can propel themselves through the water. But such a reproductive system is of no use on land, for the reproductive cells will dry out unless they are surrounded by fluid, and they can move about only in water, not in air. The earliest land plants were faced with a three-part problem: they needed a new type of reproductive system that would allow fertilization in the air; they had to be able to disseminate their reproductive cells; and they had to be able to disperse to new territories and then exploit them.

The earliest solution to the reproduction problem was solved by the evolution of spores. These microscopic bodies are small enough to be carried by the wind and have nearly watertight walls that are highly resistant to drought. But spores are produced asexually; when they finally alight, the plants they produce must then produce male and female gametes, or germ cells, if sexual reproduction is to take place. The spore method thus requires an alternation of generations between sexually and asexually produced plants. This was the method used by the late Silurian and early Devonian land plants, vanguard of the first wave of plant immigrants. It is still used today by the mosses, ferns, and liverworts. Almost all such plant forms, however, are restricted to wet or humid environments, for they depend on gametes that must be fertilized in water.

By the start of the Devonian Period, some 400 million years ago, all of the structural pieces were in place for the colonization of land. During the succeeding 40 million years the first wave covered the landscapes of the earth with the green we so freely associate with plants, but a color only lately arrived on a land surface that dates back more than 4 billion years.

The earliest known vascular plants of the early Devonian would look very strange indeed to us today. They were small, simply branched forms without true leaves—tiny plants only inches high, living in swampy areas at the edges of lakes and seas. The early land plants rapidly increased in size and complexity. With increased competition from newly evolved plants, many of the earliest land plants soon disappeared, to be replaced by species more efficient in acquiring water or nutrients, or better adapted to colonize a variety of substrates through better root systems or faster growth rates. True leaves had not yet appeared, but flattened stem surfaces provided the area necessary for photosynthesis. As the Devonian Period progressed, the rapidly

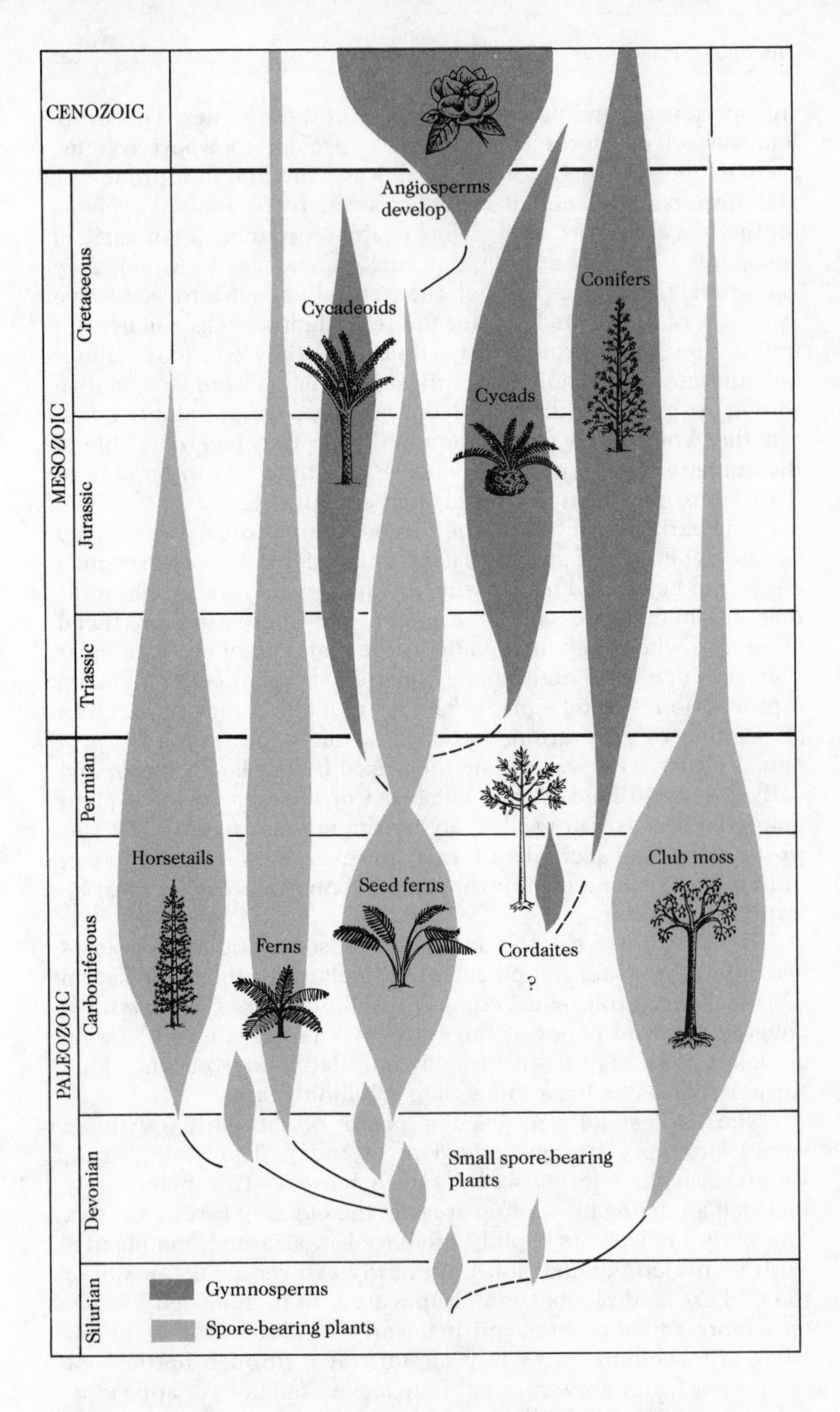

The evolutionary relationships and times of appearance of the major groups of plants.

evolving plant communities came to be dominated by groups more familiar to us, plants related to the still-living ferns, club mosses, ground pines, and horsetails. Such plants live now on the fringes of today's plant ecosystem, but during the Devonian they occupied center stage. By the late Devonian, some 360 to 380 million years ago, forests had appeared—trees forty or more feet high, with trunks as much as three feet in diameter. The plant communities, well established in many habitats, were poised for one of the great forest-building episodes in the history of the earth—the Carboniferous Period, named by the pioneering geologists of the nineteenth century for one attribute—an abundance of coal. For coal is only the compressed, metamorphosed remains of fallen trees, of ancient forests.

The first wave of plant invaders covered the earth in relatively rapid order. From the first tiny upright plants to the proud giants of the Devonian forests, the process took several tens of millions of years.

The Coal Swamps

It's an early fall day in Appalachia, and the small knot of geologists surrounding the portly gentleman spinning tales on a fallen log are enthralled. We're listening to one of the giants of North American geology, James Schopf, who dedicated his career to the study and understanding of the Paleozoic coal swamps and the plants that lived in them. At this time, in 1977, Jim is at the tail end of his long career, a kindly man with endless stories and the patience to listen to and nurture the students—and neophyte paleontology professors—drawn to him and his greatness.

We've spent the day visiting Carboniferous-aged sedimentary deposits, containing the highest-grade anthracite coal on our continent. I'm not sure what I expected to see in these Appalachian coal mines, but it wasn't at all what we did see—narrow black seams of coal cropping out of brushy hillsides, dingy beds slanting down into the ground, places where men had to lie on their backs to knock out the coal. Somehow I expected great thick deposits of coal, not these beds sometimes less than three feet thick; perhaps I thought this coal would be wrested from great open-pit mines, as it is in Wyoming, rather than from the depths of the hills around us. One thing was made abundantly clear to me—I never wanted to be a coal miner.

Jim is spinning a tale about a tiny plant he wrested from the boggy ground behind his seat. He talks around a huge chaw of tobacco as he tells us about the tiny plant he now holds, a living fossil called *Equisitum*, the horsetail. Jim suggests that it may be the oldest living creature still present on earth, a genus now composed of twenty species scattered about the world, growing mostly in ponds and ditches and marshes, any moist place. The horsetail is a plant of great antiquity, the genus dating back at least to the Triassic Period, the species perhaps all the way back to the Carboniferous Period.

We've spent the day collecting plant fossils from the coal seams, and many of them come from plants that look very much like the tiny horsetail in Jim's hands only much, much larger. I've seen many of these curious tiny plants before, as perhaps most of us have in our childhoods, when we're all natural historians and still see the plants and animals around us with open eyes and minds. Horsetails grow in swampy ground across the country, rising up each spring with their strange jointed stalks and spiky, segmented appendages. Jim asks us to imagine these plants as giants, growing in the great swamps that covered immense areas of North America and Europe some 300 million years ago. These swamps were thousands of times larger than the Everglades of Florida, dank, steaming places where great forests of horsetails, club mosses, and ferns grew swiftly to great heights and then crashed to earth to rot and be buried in the sediment. The plants were not alone in these great swamps. Giant amphibians thrashed about, seeking smaller prey, and the air was abuzz with dragonflies and other insects. Scorpions and monstrous centipedes moved about in the rotting vegetation, while on drier land the earliest reptiles began the vertebrates' conquest of land.

This riotous assemblage was the culmination of the first great plant invasion of the land: the end result of the first tentative, ground-hugging forms of the late Silurian. By the end of the Caboniferous Era the continents had moved together into a single great mass, with huge interior regions far from the moderating influence of the oceans. The ensuing Permian Period, from 300 to 250 million years ago, brought a very different climate, a time of harsh dryness, of continental glaciation on an unprecedented scale, and of continent-wide deserts. The twin killers of cold and dryness put an end to the giant swamps. A very different type of plant was needed, plants with better ability to weather bad times or to disperse to new areas. Natural selection began to work on the plants evolved during the first wave, favoring those species capable of meeting and surviving the new regime. In the sea, the greatest mass death was under way, removing forever most of the species characteristics of the Paleozoic Era. On the continents, the dominance of the amphibians was at an end, for the drying of the great coal swamps doomed them. The land vertebrates that survived were those capable of producing eggs that could hatch on land rather than only in water. The reptiles rose from the shrinking swamps to claim an arid and desolate world. Several

Common constitutents of the coal swamps were (from left to right): the Carboniferous club moss Lepidodendron, *seed ferns, and* Calamites *which was a giant sphenopsid closely related to the living horsetail plant.*

groups of plants also evolved a new body capable of reproducing in the dry, cold earth of the latest Paleozoic Era—plants that propagated with seeds. The second wave took over the earth.

The Second Wave: The Emergence of the Gymnosperms

The giant plants of the Carboniferous coal swamps of 300 million years ago were the lycopsids and sphenopsids, which survive today in the club mosses and horsetails. Both of these ancient plant groups require moisture to live, and both grow upward each spring from buried masses of rootlike structures called rhizome systems. Today they are found only in areas where the soil is soft and easily penetrated—conditions that largely disappeared from the earth when the giant coal swamps dried out, 300 million years ago. The dry, compacted soils and shifting sands of the Permian Period were unsuitable for many of the ancient plants. The result was extinction for most species of the first wave.

The successors to the floras of the Carboniferous Period did not suddenly appear and take over the landscape when the period ended. Like so many successors that rise to dominance after wholesale extinctions of more archaic organisms, they had already been part of the biota, minor elements waiting in the wings. But when the earth's climate turned cold and the swamps dried up, a drastically different flora gradually took over. This second wave was dominated by two groups that are still very much with us in our modern world: the ferns and the gymnosperms.

The ferns so familiar to us can be traced back to the Devonian Period. The earliest evolved species showed the characteristic structural innovations that we associate with this group today: lacy fronds to catch light and a complex vascular system to pass water and nutrients to all parts of the plant. The fern species that first appeared were gradually replaced in the late Carboniferous by more advanced species, the first "true" ferns, forms that are the direct ancestors of all modern fern species. They were adapted to life in the new climate, and hence were capable of surviving and reproducing in the drought and dryness of the Permian world, thanks in large part to their efficient cuticular covering and highly resistant spores. The ferns adapted rapidly to a great variety of habitats and became dominant members of the early and mid-Mesozoic floras. More than 12,000 species of ferns are still alive today, inhabiting a wide range of environments, from wet and warm to cold and dry.

The second group of plants to emerge from the floral extinctions of the late Carboniferous to attain dominance in the Mesozoic was the gymnosperms—pines, firs, and spruces, as well as rare forms such as ginkgos and monkey trees. All gymnosperms were once considered to be derived from one common plant ancestor; more recent studies, however, have shown that this group is composed of an aggregate of separately evolved lineages that share common characteristics. The ancestors of many of these groups, like those of the ferns, are found in the Devonian. They revolutionized plant life because they were the first group of plants to evolve true seeds.

Before seeds evolved, the early land plants had to alternate between sexual and asexual generations. These primitive plants reproduced and dispersed to new habitats largely by means of wind-borne spores. Most spores were so small that they were greatly affected by adverse environmental conditions. With the evolution of the seed, plants could incorporate foodstuffs with the reproductive DNA, so that the genetic information in the gamete had a better chance of surviving and being passed on to the next generation. The first seeds had no protective covering of any sort, but eventually a thick protective cover evolved to increase their survivability. With the evolution of the seed, alternating generations of sexually and asexually reproducing plants were no longer necessary. The seed-bearing plants evolved an organ called an ovule, a specialized outgrowth of the ovary that caught wind- and animal-borne pollen and protected it till it had fertilized the ovary. The ovule then produced seeds.

The earliest seed-bearing plants are found in Devonian strata; called protogymnosperms, these species rapidly diversified, and by late Carboniferous time had produced several lineages of plants that we now lump together as gymnosperms. Unlike the majority of plants of the Devonian and Carboniferous periods, these new plants no longer needed the swamps, and thrived during the cooling and drying of the Permian. They produced forests of giant trees as well as smaller shrubs, and today are still represented by the largest of all trees—the giant redwoods. Among the oldest gymnosperms were conifers, which soon diversified into pines, araucarias, firs, and ginkgos. Over 700 species of gymnosperms exist today, and in many parts of the world they still are the dominant floral elements. But in many respects their time of greatest dominance is over. Like the ferns, many still survive. But also like the ferns, they reached their zenith during the first two-thirds of the Mesozoic Era. They became the food of the dinosaurs.

The Redwoods

They stand like great, silent giants in the slanting sun. Their upper branches wave in the slight breeze 200 feet above the forest floor, green needles dusted with yellow pollen amid the ripening cones. Here, far above the forest floor, only rare insects and an occasional bird can be found among the branches. Far below, the great trees stand rooted in reddish soil, their thick bark giving off a delicious piney smell in the suffocating heat. The forest floor is in twilight; only occasional streams of light break through the canopy above. It's quiet on the floor of the forest, in the gloom of these giant trunks; the thick coat of fallen needles muffles the sighing of the wind far above. A few low bushes are interspersed among the giant redwoods, but for the most part the forest is of their making and keeping.

In the afternoon a herd of sauropods move through the forest, their long necks reaching upward for the green-laden branches of the redwoods, but little fodder is still to be found in this forest, for the constantly foraging dinosaurs have long ago stripped away the tender needles off the lower branches. The sauropods move on toward the setting sun, searching for more open terrain where food is more accessible. They pass by the small bush that grows in one of the more open glades beneath the giant redwoods, for the sauropods are used to looking upward for their food. In the now-fading afternoon light a flower is open, dazzling pink and white. A small fly, providentially carried near the bush by the capricious wind, settles on the flower, inadvertently covering itself with pollen in the process, and then heads off into the wind, carrying with it the seeds of revolution.

The Great Valley

Three in the afternoon is a hard time for the field geologist. If you're in one of the dry places of the earth, places where rocks are always best exposed, it's usually the hottest time of day. By this time I'm walking through a rocky draw in the foothills along the northwestern Sacramento Valley, surrounded by steeply inclined sedimentary strata. The low hills around me are already burned to a golden brown, the greenery of April now long gone. Ahead of me, oblivious of the heat, the time of day, and the boring rocks around us, strides one James Doyle, a professor of botany recently arrived from Michigan to begin a career at the University of California. Jim is a paleobotanist, specializing in the rise of the flowering plants; his career up to this point has been spectacular, for he has largely succeeded in working out the answer to one of the great evolutionary mysteries, a problem that vexed Darwin to frustration: Where did the flowering plants come from? It had long been known that plants bearing flowers, and the entirely new reproductive system that flowers allowed, seemingly sprang forth into the fossil record fully formed, with no record of direct ancestors, sometime early in the Cretaceous Period. Many paleobotanists had tackled this problem, and most had concluded that flowering plants, termed angiosperms by the botanists, had originated much earlier than their first fossils indicated. But Jim Doyle has succeeded in showing that the angiosperms arose quickly, through an explosive burst of evolution during the early Cretaceous, about 100 million years ago. He is now trying to examine this radiation of the first flowering plants on the California coast, so he can compare his findings here with his discoveries on the East Coast.

We drove northward from Sacramento on I-5 in the morning. The wide expanse of the Sacramento Valley produces almost an eerie feeling; you could swear you were in the Midwest, not California, for the Great Valley is so wide and flat that it speaks of Nebraska or Kansas. Only the distant Sierra peaks give away our true position as we pass mile after mile of rice fields and orchards.

After two hours of freeway passage we finally arrived at Corning ("Olive Capital of the World!") and headed west. The flatness of the valley finally gave way to low undulations, then to steeper hills, and finally culminated in a low mountain range, the western side of the valley. From our vantage point the view is breathtaking. To the south, the valley seems to stretch to infinity; it's many hundreds of miles to the southern limit, near Bakersfield. We can barely see the Sierra side, nearly fifty miles away, to the east.

The foothills to the west of the valley are made of one of the thickest piles of sedimentary rock known in the world. In the Cretaceous Period, long before the Sierras arose to their current glory, the Great Valley of California was a giant seaway. Sediment from nearby hills eroded into this sea, rapidly filling the basin. The weight of the accumulating sediment caused the earth's crust to subside, making room for yet more sediment. Eventually, near the end of the Cretaceous Period, the rapid sedimentation in this region stopped, but not before the sand, gravel, and mud had accumulated to a depth of five miles on an immense sea floor the length of California.

We're currently somewhere in the middle of this great pile of rock, now thrust up from its deep oceanic resting place by the gigantic tectonic forces that created the Sierra Nevada range. Unfortunately, the very thickness of the Cretaceous-aged sediment that makes up the western side of the Great Valley presents many difficulties. Jim is hoping to collect samples from one very narrow time interval in this great accumulation of rock. In Michigan, sediment thicknesses are usually measured in tens of meters, not tens of kilometers. He blithely assumes that I can march up to the rocks, put my finger on the magic spot, and declare the age of the rock.

The strata here are nearly vertical, clearly having been uplifted from the horizontal since they were deposited, over 100 million years ago. They are also nasty rocks called turbidites, and, because of their ubiquitous presence, they must be California's state rock. This type of sedimentary rock is deposited in deep marine basins, the end result of submarine landslides. Each turbidite deposit usually measures from several inches to several feet thick at most, but they have piled up, one layer atop another, in the thousands to millions over the eons. They almost never contain fossils, and most of the rare fossils they do contain have been transported from shallower water and fragmented by the catastrophic underwater flows.

Jim has come to find fossil pollen, which he hopes was carried into these marine strata during the Age of Dinosaurs. But secretly he has a greater hope: to find leaves of the earliest angiosperms. He needs rocks of a particular age, called the Aptian Age of the Cretaceous Period, for it was during or immediately before this narrow interval of time that the flowering plants burst into being and very rapidly began to cover the earth. It was one of the most spectacular adaptive radiations in the long history of the earth; within about 30 million years of their first appearance, the flowering plants dominated the floras of the earth, as they still do today, pushing aside the once-dominant gymnosperms.

We work our way farther up the dusty draw, climbing through time as we go. As we pass around a rocky ledge we come across the long-dead body of a cow, its bones now bleached by the sun. I can't help thinking of many more pleasant outcrops in Northern California we could be visiting. Jim keeps asking if we've reached the right rocks; like so many paleobotonists, he has great confidence in a "specialist's" ability to gauge the ages of rocks. But since I haven't seen a fossil in the last half hour, I haven't the faintest idea when these rocks may have been deposited. The best I can say is that we're in the ball park of the rocks he seeks, so just go ahead and sample everything.

In the distance I see the end of the long outcrop and spy a solitary figure trudging toward us in the hot sun. His figure shimmers in the heat. We finally meet and exchange greetings. Jim has his eyes on the outcrop as I pause to have a word with the stranger. He's dirty and dusty, so he looks much like us. He spies my hammer and breaks into a grin. "You fellas looking for gold?" I've already noted the pan strapped to his pack; he's one of the legion of prospectors still panning the rivers and creeks of Northern California, looking for show and gold flour, or the elusive nugget that escaped the forty-niners and their successors. No, I reply, we're looking for fossil plants. The stranger looks at me in astonishment. "Fossils, in this heat? You guys are crazy!" I secretly agree as I trudge on after Jim, now well along up the outcrop.

We finally call it a day after five o'clock, our packs bulging with rock samples, all bagged in cloth. We're stretched out under one of the giant oak trees that dominate these foothills, enjoying the shade before we hike back to the car with our dubious treasures. I feel as I usually do after a day in the valley: I'm parched and sunburned, and my feet hurt. We stare out over the valley spread out before us, at the endless expanse of irrigated fields, for the Great Valley is surely one of the world's richest agricultural areas; it produces as much as a quarter of all fruits and vegetables grown in the United States. I see endless miles of trees and vegetables, distant clover fields, and barley, corn, and wheat, a treasure of greenness. Jim is staring out over this world of plants, too. But he sees plants with eyes much different from mine, and he points out a simple fact: every plant we see, from the strawlike grass around us to the great oak over our heads to the emerald lushness of the valley below, every plant for miles around is an angiosperm, a flowering plant.

We haven't found a single fossil leaf or any piece of rock that contains anything identifiable as a plant. But the treasures Jim seeks are microscopic; he hopes his samples contain fossil pollen, microscopic evidence of the greatest of the plant revolutions. Sometime in the early Cretaceous a miracle occurred: the first flower bloomed, and forever changed the face of the earth. That first flower signaled the onset of the third wave.

The Third Wave: The Rise of the Angiosperms

By early April the incessant rains of the Pacific Northwest at least warm a bit and lead to speculation that perhaps spring may come after all—a feeling substantiated by the lengthening days. But, ever distrustful, we look for the flowering of trees as proof of the changing of the seasons. The crocuses have long since bloomed and the daffodils are now in full glory, but somehow it's the trees, not the flowering bulbs, that testify that winter has lost its grip, and that once again spring has brought its gift of renewal. The dogwoods begin to swell, and the earliest rhododendrons and flowering cherries break out into sudden splendor. But of all of the signs of spring, one plant gives greatest hope. These plants are almost ridiculous in their flair for the dramatic, putting forth gigantic flowers that precede the green of leaves—huge trumpets of bloom ranging from splashy white to bright pink. All over the Northwest, and in most other parts of the temperature to semitropical world as well, the magnolias begin the parade of spring. They range from low bushes to large trees, and it doesn't take a botanist to know that there's something special about them. The magnolias have reason to burst forth first among the trees, for over 100 million years ago they, or species much like them, were among the first plants to produce flowers. I find it pleasing that the onset of the flowering plants was marked by the most magnificent flowers of all.

That the angiosperms are the dominant plants of our world is undisputed. They grow in a wider variety of habitats, show more variation, and, perhaps most important, have the largest number of species of any plant group living or dead. Botanists have identified more than 257,000 species of angiosperms—five times as many species as all other plant groups combined. They are the masters of this world, and their rise to dominance took place with breathtaking rapidity.

No single characteristic can be used to define angiosperms, for many features have contributed to their success. The most obvious characteristic is the presence of flowers, the hallmark of a reproductive system far more efficient than that of any nonflowering plant. But other features are important as well—the veined leaves, the highly efficient vascular transport system, an improved support system, and even the habit of dropping their leaves, as many angiosperm species do.

The origin of the angiosperms and the nature of their earliest evolutionary pathways and dispersal have been the subjects of lively and often acrimonious debate for decades. Darwin called the origin of the angiosperms "an abominable mystery," no doubt because of their seemingly sudden appearance in the fossil record of the Cretaceous Period strata. Darwin could not countenance the appearance of complex organisms without fossil evidence of a long period of evolution, and the seemingly instantaneous appearance of complex angiosperm leaves in middle-Cretaceous strata in England and many other parts of the world drove him to distraction. In a relatively short time the world of 100 million years ago was taken over by the flowering plants. By late Cretaceous time this third wave of plant invaders had swept the earlier floras away.

The apparent rapidity with which the angiosperms acquired dominance over the gymnosperms led many paleobotanists to speculate that flowering plants had evolved long before they actually appeared in the fossil record. It's easy to see how this notion originated, for the breaktaking takeover of the world by the flowering plants does smack of a well-planned invasion. Proponents of this idea speculated that the angiosperms evolved during pre-Cretaceous times in some unknown "homeland," an environment that left no trace of itself in the fossil record. Perhaps the homeland was a high-latitude region, even the Arctic or Antarctic; or perhaps it was a more temperate region but high in the mountains or in another environment where little sedimentation takes place, and hence leaves no record of any creatures that may have lived there. Certainly there is abundant precedence for such a phenomenon. Many creatures, such as the coelacanth, live in areas that preserve no fossils, and so exist for long periods of time undocumented by the fossil record. But the proponents of the homeland hypothesis went further: they suggested that angiosperms underwent a long period of evolution and diversification in the homeland, so that when they finally burst forth from their mysterious base, they did so already diversified and preadapted for the many environments in which they are rather suddenly found in the middle of the Cretaceous Period. This scenario conjures up wonderful visions: imagine long lines of oaks, and maples and beeches and all the rest of the familiar angiosperm trees, as well as innumerable shrubs and weeds, all patiently waiting for the invasion order, marshaling their seeds for the moment of conflict, thinking last thoughts, perhaps, of that shady grove of their saplinghood; and then the order to march, trees to the forests, grasses to the plains, pond lilies and rushes to the swamps, each preadapted group bravely flinging itself into conflict with the entrenched armies of ferns and horsetails, pines and cycads, taking fearful losses at first but soon putting the armies of the gymnosperms to rout.

Two great paleobotanists put this fable to rout. In the late 1960s and early 1970s, Jim Doyle and Leo Hickey showed conclusively that angiosperms first evolved early in the Cretaceous period, not before, and diversified rapidly among the preexisting flora, rather than invading from some unknown homeland. Doyle and Hickey came to these conclusions by conducting detailed studies of the fossil floras collected from early Cretaceous-aged sedimentary strata on the east coast of the United States. These rocks, called the Potomac Group, stretch along a wide belt from Pennsylvania through Maryland and into Virginia. Exposed on the rivers of the region—the Potomac, the James, the Appomattox—the Potomac Group consists of a wide variety of sandstone, shale, and gravel deposited in ancient rivers, streams, ponds, and floodplains, as well as deltas and alluvial fans. These deposits of early Cretaceous age contain leaf fossils as well as fossil pollen. It's the pollen that essentially served to pinpoint the origin of the angiosperms, for, unlike leaves, which are preserved only under rare circumstances and in very special places, pollens are hardy and are preserved in a wide variety of sedimentary environments. Doyle and Hickey showed that the earliest angiosperm pollen was very primitive in the oldest Potomac Group Cretaceous strata, and hence had not undergone some long evolution before the Cretaceous Period. Their studies also showed that the pollen had come from plants that lived in or near the areas where the Potomac sedimentary strata were deposited—not in some faraway homeland.

Doyle and Hickey's studies yielded fascinating insights into the earliest evolution of the flowering plants. The oldest known angiosperms appear to have been small, weedy plants, in many ways more similar to living magnolias rather than to large trees. These earliest flowering plants also seem to have lived near water, perhaps by streams or ponds, and were adapted to exploit disturbed environments rather than to live in stable ecosystems. They thus may have been forms that specialized in colonizing such environments as newly emerged floodplains. From this starting point newly evolving angiosperm species seem to have invaded the forests as understory thickets and shrubs. Within several million years of their first appearance they left evidence of their presence in a wide range of environments, creeping north and south toward the poles. The rate of their diversification was extraordinary. At the start of the Cretaceous Period, about 125 million years ago, about 30 percent of the earth's plant species were ferns, another 30 percent cycads and other tree and seed ferns, and the rest gymnosperms, as well as more archaic forms such as lycopsids and sphenopsids. Ten million years later, when the first angiosperm pollens appear, these percentages were about the same. But 10 million years after that the entire world had changed; angiosperms were taking over the myriad environments of the land. By mid-Cretaceous time, 75 million years ago, more than two-thirds of all plants that became fossilized were flowering plants. The angiosperms had conquered the earth in a remarkably short time.

The consequences of the angiosperm invasion extended far beyond the makeup of the terrestrial floras; the animals of the earth were severely affected as well. Two groups of animals were most intimately related to the plant ecosystems: the terrestrial herbivores, which depended on the land plants for food, and the terrestrial insects, which in large part were instrumental in the reproductive success of the flowering plants.

Dinosaurs and Land Plants

By late Jurassic time, between 150 and 125 million years ago, the assemblage of creatures known to all of us as dinosaurs had reached their peak in numbers, diversity, and especially size. With their tiny heads at the ends of long necks, their huge, barrel-shaped bodies, and their long, whiplike tails, the sauropod dinosaurs evolved into the largest creatures ever to live on land. And not only were they big; there is evidence to suggest that there were a lot of them, as well.

Some of the best-known dinosaur-bearing deposits of late Jurassic age come from the American Southwest, where sandstone and mudstone strata of the Morrison Formation are exposed. The Morrison Formation contains the largest accumulation of dinosaur bones of any age yet discovered on earth, and gives us valuable information not only about the species of dinosaurs present on earth when these river and floodplain deposits were accumulating, but about the relative numbers of the various inhabitants of that ancient world as well. In a fascinating study the paleontologists M. Coe, D. Dilcher, J. Farlow, D. Jarzen, and D. Russell analyzed the Morrison dinosaur assemblage and came to a startling conclusion. They compared the Jurassic Morrison dinosaur assemblage with the closest possible living analog—an undisturbed African large-mammal ecosystem. The large land mammals of the African savannas—elephants, hippos, giraffes, rhinoceroses, the larger antelope species—are found in populations of varying size across their range. In those areas where humans have disturbed the environment, their populations are small. In the relatively undisturbed regions, such as the larger animal reserves, and in areas where rainfall is abundant, relatively accurate estimates of populations per unit of territory can be made. These estimates can yield the average biomass that the ecosystem can sustain, as well as the weight of plant material consumed by the herbivores of the area in any given time period. Such figures have been gathered for the Amboseli Plain of southern Kenya. In this region cattle are the most common of the larger mammals, with an average of ten found on each square kilometer, for a total biomass of nearly 5000 pounds. Elephants, although far less numerous (only, on the average, one for every *four* square kilometers), turn out to have a higher biomass, because the average weight of an elephant is so much greater than that of a cow. It was calculated that one square kilometer of territory in this region of Africa supports slightly fewer than twenty-four animals of all types, with a total average biomass of about 10,000 pounds. Using some very clever computations based on the average size of a leg bone, Coe and his colleagues estimated the average weight of a dinosaur; and by laboriously counting the number of individual dinosaur fossils found in Morrison Formation deposits, they arrived at rough estimates of the number of dinosaurs that might have lived or been supported on each square kilometer of land back in the late Jurassic on the plain that yielded the Morrison Formation dinosaurs. They found that each square kilometer would have had about the same number of dinosaurs as of mammals, but because each dinosaur weighed much more—*much* more—than the average mammal, the dinosaurian biomass per square kilometer would have been an incredible 200,000 pounds: twenty times the value for a comparable area of the modern African savanna.

If these figures are anywhere near accurate, the implications are staggering: imagine two or three city blocks (about a square kilometer) and put in that area three giant sauropods the size of a brontosaurus or diplodocus, one or two stegosauruses, and a very hungry carnivorous allosaurus. This estimate implies that the world was covered with dinosaurs, awash with dinosaurs everywhere, in vast herds trampling the earth and eating everything in sight. Consider the plight of the plants when this vast, scaly, hungry reptilian horde covered the earth. It was a world so different from our own that we can't begin to imagine it.

Ecologists have a wonderful euphemism for environments that are constantly subjected to change: they call them "disturbed" environments. With all those giant dinosaurs crashing around, stripping leaves off plants and tromping the shrubbery flat, the entire world must have been disturbed. The sauropods, such as *Brontosaurus* (*Apatosaurus* for the purists), *Brachiosaurus*, *Diplodocus*, and the new "giant" sauropods, such as *Ultrasaurus* and *Amphicoelias*, were the most common dinosaurs of the time, giant plant-eating machines with necks that allowed them to browse comfortably on tree limbs as much as twenty-five feet above the ground, and even higher if they rose up on their hind legs. The effect on the floras of the time must have been devastating. The landscape would have been largely open, without dense forests, owing in large measure to the constant foraging and feeding of these giant herbivores. The sauropods had tiny heads with simple teeth, which seem to have been adapted for stripping foliage and twigs from plants, and probably were not used for chewing. The huge sauropods probably lived in open country rather than in dense jungle or forest. Pine needles may have made up a large part of their diet, and because their metabolism was very slow, the sauropods may have used a long, slow fermentation process, keeping plant material in their stomachs far longer than modern herbivore mammals do. One vertebrate paleontologist has whimsically described the herbivorous dinosaurs as "walking compost piles."

It's in the context of this incredible world that we must view the evolution of the earliest angiosperms. In a world where trees were constantly being stripped of their leaves and branches by the sauropod herds, where dense forest was rare, and where the ability to take root and grow rapidly would be a huge advantage —in such a world it made sense to be an angiosperm. The angiosperms' seeds carried more food material than the gymnosperms' seeds did, so the angiosperm seedlings were larger and hardier and grew faster. Their seeds also had a protective covering, and were probably capable of surviving for some time in the gut of a large dinosaur; in this fashion the earliest angiosperms may have been rapidly dispersed around the globe. We know that only 10 million years elapsed from the time of the first angiosperm fossils of the early Cretaceous until these plants were distributed throughout the world, and it's highly probable that this progressive dispersion can be credited to the dinosaurs.

By the end of the Cretaceous Period, some 66 million years ago (about 40 million years after the flowering plants first appeared), the sauropods had all but disappeared, replaced by the duck-billed and cerotopsian dinosaurs—*Triceratops* and its cousins. Although these creatures were still behemoths, they were much smaller than the sauropods, and they appear to have fed in different ways as well. The herbivorous dinosaurs of the late Cretaceous fed on plants low to the ground, rather than on the branches and tops of tall trees; their teeth show that they were capable of chewing their food, so their digestive processes may have differed from those of the sauropods as well. And finally, their food resources must have been markedly different, for the late-Cretaceous world was dominated by a flora not so dissimilar to ours: a flora dominated by angiosperms.

Were any of these great changes in the herbivorous dinosaurs brought about by the changes in the flora of the times? Many paleontologists think so. By latest Cretaceous time the floras of the world differed radically from the late-Jurassic floras. Angiosperms had taken over the earth, and many new types of animals were rapidly evolving to exploit this new type of plant. The fossil evidence indicates a trend toward increasing size of seeds and fruits during the late Cretaceous. The earliest angiosperms had very small seeds, but increasing numbers of angiosperm taxa found it advantageous to add food material to their seeds, and thus to increase their size. This new source of energy provoked the evolution of creatures specialized to exploit it. Seed eating, however, is a habit that favors small animals. The number of very small dinosaurs increased toward the end of the Cretaceous, as did the numbers of some other animals that may have found it advantageous to eat seeds and fruits: the early mammals.

The rise of the angiosperms during the Cretaceous Period also changed the very nature of the forests. During latest Jurassic and earliest Cretaceous times there may have been few large forests as we now know them; most of the large gymnosperms and cycads of 150 million years ago were scattered across the landscape, so they had few opportunities to form dense stands of trees and canopy. The actual number of tree species was much smaller than it is today. The initial evolution of the angiosperms during the early Cretaceous Period did little to change the nature of flora, at least at first, because it took many millions of years for large angiosperm trees to evolve. Gradually, however, larger angiosperm plants did begin to appear, and by the last of the Cretaceous Period, some 75 to 65 million years ago, true forests of flowering plants had displaced many of the areas previously populated by the gymnosperm-dominated savannas. These new forests produced a thick jungle with an enclosed canopy, no place for tall, long-necked dinosaurs. With the passing of the gymnosperm-dominated world, the sauropod dinosaurs also passed from the earth. The dinosaurs that remained were shorter, grazers rather than tree browsers. An era had ended, and the earth waited for the final curtain to fall on the dinosaurs.

The Fern Spike

In 1980, as we have seen, Luis and Walter Alvarez and their colleagues proposed that the Cretaceous ended not with a whimper but with a bang: the finding of extraordinarily high concentrations of iridium in the clays of a Cretaceous-Tertiary boundary site in Italy persuaded them that a huge asteroid had hit the earth with such force that the environmental effects led to wide-scale extinction of plants and animals, the most famous of the casualties being the dinosaurs. Luis Alvarez was a Nobel laureate, so the hypothesis seemed to merit investigation. Several teams of geologists and paleontologists set out to study well-exposed sedimentary rock sections in the western United States. As we know, they too found high levels of iridium in the clay boundary layers, just as the Alvarez group had predicted. But because the North American strata were not marine sedimentary deposits, as those described by the Alvarez group were, these investigators made a totally unexpected finding as well: the pollen record across the Cretaceous-Tertiary boundary showed a profound change. The paleobotanist R. Tschudy found that the angiosperms of the latest Cretaceous were suddenly, almost totally replaced by ferns. He called this change in pollen the fern spike.

The finding of the fern spike was rapidly confirmed at other nonmarine Cretaceous-Tertiary boundary sections. Paleobotanists soon reconstructed the floral makeup of the latest Cretaceous world of western North America immediately before the boundary clays were deposited. The landscape they evoked was semiarid, dotted with many species of angiosperms and a few gymnosperms. Most of these plants had small leaves and thick cuticles, evidence that the climate was warm and dry. Then suddenly the entire flora was disrupted. Virtually every plant disappeared, to be replaced by an assemblage of ferns with little diversity. Gradually, over about 1.5 million years, the higher plants invaded the region again, displacing the ferns. But the type of plants found above the Cretaceous-Tertiary boundary are entirely different from those below the boundary; they have broad leaves and seem adapted to a cooler, wetter environment.

The discovery of the fern spike became one of the strongest bits of evidence that catastrophe visited the earth some 66 million years ago and profoundly affected plants and animals then living on the earth. Ferns are often the first plants to appear after a catastrophe has destroyed a flora, as after a forest fire or a volcanic eruption. The ferns, with their wind-blown spores, can quickly colonize a devastated area; more slow-growing plants will invade the area later and gradually displace the ferns. This phenomenon seems to have occurred at the end of the Cretaceous. The widespread forests and floras of the latest Cretaceous world were suddenly displaced by ferns. There seems to be little alternative to the conclusion that catastrophe did indeed visit the earth.

The very climate was changed by the event that ended the Cretaceous. An innovation of the post-Cretaceous world was the evolution of grasses. As the climate of the earth gradually cooled to its present-day levels, vast grasslands covered large portions of the earth. After the dinosaurs were gone, the world was inhabited by no large herbivores for several millions of years. But nature abhors a vacuum, and the abundance of plant life reestablished in the post-Cretaceous world was too good a resource not to be exploited. In short order the few small, shivering rats that survived the apocalypse gave rise to the wide spectrum of mammals that now populate the earth. We are here because those rats survived.

8

Out of the Ooze
The Lobe-fins

Rendezvous with a Fossil

I first encountered a coelacanthid fish—or at least what had once been part of one—in my twenty-third year. I had just started my doctoral dissertation, an attempt (doomed, as it turned out) to demonstrate that ammonite fossils showed distinct distribution patterns within similarly aged strata on Vancouver Island in Canada. In this region the Cretaceous-aged strata I was investigating can be found only in river canyons and on the coasts of islands; everywhere else, rain forest obscures the land. Because the fossil-bearing rock is soft shale, the creeks and rivers have cut gorges so deep and steep that the waters below are all but inaccessible. Over the weeks of that summer I walked the land, chasing ephemeral and long-extinct logging roads that my obsolete topographic maps promised would lead me to the river bottom.

One of the deepest of the canyons has been cut by a stream called the Trent River. I walk the top of this deep canyon for many miles, looking down the fern-covered sides at the distant black shales glistening in the river bottom so far below. On a cold, overcast day I find a way into this canyon, and as I descend into the gorge I drop down in time. When I finally reach the river I'm standing on rocks that were deposited at the bottom of a muddy sea nearly 80 million years ago.

Descending these deep gorges is very much like taking a deep scuba dive. Once you reach the bottom, a moment of calm arrives, and euphoria as well. But these joys are always tempered with the knowledge that bottom time is borrowed time. Here, as at the bottom of the sea, my stay will necessarily be short, and I have much work to do. I stare at the shales, searching for the telltale spirals or glints of color that reveal the presence of ammonite fossils. The calls of the birds, the rush of the river, the gentle sighing wind in the trees, all fall away as my eyes scan the shales, moving reluctantly on from joining patterns and cracks and other false signals of ancient life. And gradually the rest of the world falls away and the fossils begin to appear. Almost shyly the most obvious forms call my glance, and then, as my concentration increases, more subtle shapes emerge from their story cover.

The Trent River strata hold great treasures. These strata must have been deposited on a deep, quiet bottom, for no aspect of the rock suggests waves or currents. The extremely fine-grained shales were deposited far from shore, for not even rare sand grains can be found among the fine clays of this bottom. On my knees now, oblivious of the fine rain falling from the strip of sky visible between the high walls of the canyon, I begin collecting these last coins of a once flourishing empire.

My first impression is that the ancient sea bottom now preserved as shale is littered with white pencils. Innumerable white cones gleam dimly from the black strata around me, the shells of straight ammonites called *Baculites*. These creatures lived in unknowable numbers near the end of the Age of Dinosaurs, the straight shell portions lending neutral buoyancy to the creature that lived within the shell. Without the shell they probably looked much like small squids, but this can be only supposition, for they died out 65 million years ago, in the cataclysm that ended the Mesozoic Era. I have come to Trent River specifically to collect these fossils, and even at this time I know I will spend my life studying them.

With swift strokes of hammer and chisel the ammonites are removed from their stony graves to be carefully labeled and wrapped. Scattered among the numerous baculitid ammonites I begin to see other fossils: large, spiny snail shells and other ammonites as well. Some shells are flattened spirals, like those of the nautiluses; others, such as the *Baculites*, are uncoiled, some with the most bizarre shapes. I marvel at even rarer treasures, for the fine muddy bottom of this ancient seabed contains the carapaces

The Trent River Valley, lined by Cretaceous-aged shales.

of long-dead shrimp, some with their legs and antennae clearly visible. Most curious of all are innumerable bluish circles, something like coins, randomly scattered across the strata. I collect a few and wander from my workplace near the center of the river to the sides of the canyon, hungrily snagging purple huckleberries as I go, lost in thought. I've never before seen these round, flattened impressions, each the size of a quarter. I move to a slightly higher bench of the river and find more ammonites.

As the day goes by my rucksack grows heavy, and I begin to stare at the dark gray walls I must yet climb, thinking of the climb and the long logging road back to my car. Supper will have to be cooked before the dubious pleasures of my tent and a surely damp sleeping bag. My mind keeps returning to the puzzle of the curious round fossils littering the stratal surfaces. They are strangely familiar, but not really. Could they be the flattened remains of ammonite jaws? Ammonites, after all, had beaklike jaws not unlike those of the modern-day octopus. Perhaps if such jaws were flattened, they'd be preserved as the rounded structures I've been gathering. But such a conjecture doesn't readily stand up to serious thought, for there are far fewer ammonites here than these strange fossils.

A fine rainy mist begins again, and all thoughts of fossils fade before the immediate task of getting out of the canyon. I'm frozen, even on this July day, and I'm not looking forward to my climb and the hike back. I adjust belts and pack and begin my trek, all thoughts of the strange fossils littering this ancient sea bottom now put aside.

Several weeks have passed since my climb to the bottom of the Trent River Canyon. I am far from that place geographically and even farther climatically. The soft gray mist of Vancouver Island is now replaced by the exuberant sunshine of the Bay Area on a fine late-summer day, the smell of Canadian conifers replaced by the alkaloid fragrance of eucalyptus. I'm in Berkeley, here to visit the museum collections housed in the Department of Paleontology of the University of California. I've brought several of my Vancouver Island ammonites, and, on a whim, several of the coinlike fossils as well.

The man in charge of the Cal-Berkeley paleontology collections is Joe Peck, a retired marine officer. Joe is tall, square-jawed, and ramrod straight, with a voice admirably suited to his former calling. Military and paleontological memorabilia are scattered about his office in equal measure. The enthusiasm of his welcome seems to be tempered by the length of the ponytail sprouting from the back of my head. It is, after all, 1973.

Joe and I chat about this and that before I go off to immerse myself in the extensive ammonite collections. I tell him of the Trent River fauna and the odd round fossils I found there. I bring out one of them and carefully unwrap it. Joe picks up the fossil and asks me what I think it is. I give him my best guess, that it's the preserved remains of a cephalopod jaw element. Joe fingers the curious object, peers myopically at it through his jeweler's loup, and gives me his diagnosis. "They are fish scales, without a doubt. And judging from their size, they came from a very large fish. The ridges and shape of these scales give them away. There is only one fish they could have come from—a coelacanth."

I'm astonished. Of course I know that living coelacanths have been discovered in this century, but their last fossil appearance dates back to the end of the Mesozoic. I think once again of the fossil assemblage in the Trent River Canyon. In my mind I journey back through time, to about 80 million years ago, to a deep muddy bottom, far below the waves: a quiet, warm bottom, rich in life. Shrimp scuttle over the bottom, as do crabs and hermit crabs. Ammonites slowly browse the surface of the muddy bottom, balloonists on a slow joyride over the hummocked fields of mud. And slowly looming through the dark an ominous shape emerges with gleaming yellow eyes, giant scales, and the strange lobed fins characteristic of its tribe: a coelacanth, voracious in its hunger, attacking the larger shrimp as a host of smaller fish flee in all directions. Here and there on the surface of the sediment are the scales of another coelacanth, perhaps killed by one of the giant predators of these Mesozoic waters, a lizardlike mosasaur. Joe Peck rouses me from my reverie, a quizzical look on his face. I know what he's thinking, with this strange longhair in front of him. "Mind if I keep these?" Information is never free. The fossils are his.

The large scales still litter the strata at the bottom of the Trent River Canyon. As the seasons turn on Vancouver Island, thousands erode out of the shale banks to disintegrate in the rain or be carried away by the swift river. Their numbers are steadily diminishing, and there's no renewal of this fossil resource. The last coelacanth skeletons known in the fossil record come from the middle part of the Cretaceous Period, from rocks about 100 million years old. The scales from Trent River are 20 million years younger. If they did indeed come from a coelacanth, they represent the last known fossil occurrence of this group. But not the last known coelacanth.

Landfall

A Devonian Period night, 400 million years ago. The large scorpion is motionless as it rests on the sandy lakeshore, its brownish carapace glistening in the full moon. The night air is thick, oppressively hot. The only sound is the gentle sighing of the wind in the spiky low plants lining the shore, for with the exception of the scorpions and primitive insects, the land is largely devoid of animal life. The nearby lake, in contrast, writhes with life, its oily surface roiled by the hunters and the hunted, for the giant moon has created treacherous daylight out of the normal haven of nighttime darkness. The lake teeming with animal life, the land curiously barren, until now the only sanctuary of invertebrate phyla in their efforts to escape the voracious, multiplying, all-powerful vertebrates.

A small insect clambers by the motionless scorpion, oblivious of the large pincers. The motion of the insect is registered in the scorpion's nervous system; claws flash out and the squirming beetle is brought to the waiting jaws. The scorpion quickly finishes its meal and then rotates slightly, its numerous eyes searching for further movement on the nearby sand. Several feet away, in the shallows of the lake, many pairs of yellow eyes, gleaming in the moonlight, note the motion on the shore. A dark shape emerges from the lake, and in a sudden rush a large fish is upon the scorpion, its needlelike teeth crushing the predator-turned-prey. In its death throes the scorpion repeatedly tries to sting the fish with its poison-tipped tail, but cannot penetrate the thick scales and bony armor. The fish rests on the sand after wolfing down the scorpion. Its body is heavy but the four squat limbs at its sides offer sufficient support to allow breathing. Now completely out of the water, the fish stares inland, searching for other prey. Seeing nothing, the fish slowly, awkwardly turns back toward the lake, legs churning and tail flailing in the sand. Back into the shallows it goes. When the water is deep enough it turns, more efficiently now, and positions itself so that only its eyes are out of water, joining the thousands of others of its kind, all watching the shoreline, looking at the land.

The transition from sea to land by our vertebrate ancestors was one of the most significant events in the history of life. At some moment a vertebrate creature, a member of our phylum, completely left the water of its own volition for the first time. It may have ventured forth for food or to escape some predator, or perhaps its water hole dried up. But there was a first time. And in my opinion, the first step of a man on the moon was nothing in comparison with the first step of our vertebrate ancestor on the land.

We still know little about these earliest amphibians. They first ventured ashore sometime during the Devonian Period, a time interval that began about 400 million years ago and ended about 360 million years ago. The fossils that are universally accepted as the first amphibians were found in rocky outcroppings in eastern Greenland. These first vertebrate colonists of the land were not small, timid creatures fleeing for their lives; they were large in relation to other vertebrates of the time, about three feet long. They had large, flat heads with big bulbous eyes, and were equipped with sharply pointed teeth that most assuredly were not used to eat vegetation; our earliest land-dwelling ancestors were exclusively carnivores. They were thick of body and were probably very clumsy on land, moving about on four squat limbs splayed out to the sides and dragging a thick tail. In many respects they looked more like fish than creatures adapted for land, and indeed they may have spent far more time in water than on land, coming out perhaps only to feed, or to escape muddy streams or ponds during dry seasons. But our ancestors they were, for our skull and skeletal patterns, although markedly modified by millions of years of evolution, can be traced back to corresponding structures in these earliest amphibians. It appears that all land-dwelling vertebrates of today can be traced back to this single ancestor, this first colonist of the land.

Paleontologists are in rare agreement that this first amphibian, named *Ichthyostega*, represents the single stem that gave rise to all subsequent land vertebrates. But there is much less consensus about the identity of the fish species that gave rise to *Ichthyostega*. Paleontologists are eager to pinpoint the immediate ancestor of the first amphibia, for only when we know that can we know how and perhaps why the transition from water life to land life was made. It is agreed that the first amphibian came from a stock of fishes that had peculiar lobed fins. The problem is that three separate stocks of fish with such fins existed about the right time.

The Devonian Period may be most important to us because it was the time when our amphibian ancestors made landfall. But the Devonian is better known among paleontologists as the Age of Fishes, for it was a time of rapid diversification of many lineages of both marine and freshwater fish. During this busy time there were five great stocks of fish and several minor ones. What an unbelievable scuba dive it would be to enter a Devonian sea or lake, for few of the fish would be recognizable to us. Perhaps most foreign would be the heavily armored jawless fishes known as ostracoderms. These fish, mainly small, flattened bottom dwellers, are represented today only by the hagfishes and lampreys. More impressive and, because of their sometimes gigantic size, infinitely more dangerous to our time-traveling scuba diver would be the first jawed fishes, the placoderms, now completely extinct. Surely also of interest to our now-terrified observer would be the first sharks, even back then predatory creatures. Also fairly recognizable would be the primitive bony fish, at that time newly evolved. Although most of the Devonian bony fish might seem peculiar to us, encased as they were in thick scales and body armor, for the most part they had the familiar rayed fins—a single dorsal fin and paired pectoral and pelvic fins. A fisherman who caught one of these early fish today might remark at the scales but not at the fins. But a second group of bony fish typical of those long-ago times might not seem so familiar. Some of the bony fishes of the Devonian Period had very different sorts of fins, fins that sprouted from fleshy lobes, fins that look almost like toeless legs. And these strange fish had two dorsal fins instead of the single large one we have come to expect.

The two great groups of the bony fishes are called the actinopterygians, or ray fins, which comprise the stock of fish most common on earth today, and the sarcopterygians, or lobe-fins. It's the latter group that has two dorsal fins and paired pectoral and pelvic fins. Sometime in the middle part of the Devonian Period some species of freshwater lobe-finned fish evolved legs and commenced the conquest of the land as *Ichthyostega*.

The evolution of life has often been illustrated by a tree-shaped structure, with newly evolved groups branching off the trunk. Some branches grow and give rise to many new branches. Others live for a time and then wither and die, even if they have given rise to still viable lineages. So it was with the lobe-fins. Their greatest claim to fame was that a single Devonian species of their group gave rise to the first true land-living vertebrates. The transition surely was not accomplished in a single step, from one species to the next, or even perhaps in a dozen steps. But some creature did finally emerge from the waters of some Devonian lake to stand on dry land for the first time. And that triumphant creature had evolved from the lobe-fins.

The first amphibians came from one of the three lineages of lobe-fins in the Devonian sea and lakes; exactly which of these three actually gave rise to the first land vertebrates still provokes debate among paleontologists. It may have been the lungfishes, a specialized group of fish that still live today. But these fishes show numerous adaptations for the very specialized environments in which they live. Their few individuals and few species lead a solitary existence in isolated ponds and streams in Africa and Australia; the lungfishes are considered the least likely candidate. The second group, the rhipidistians, are thought to be much more likely to have given rise to the amphibians, for they show fin and skull structures very similar to the heads and limbs of the first amphibians. But if they were the stem group that gave rise to all subsequent land-living vertebrates, it was their sole mark of distinction, for all the rhipidistians were gone by the close of the Paleozoic Era.

The final group of lobe-fins, the coelacanths, could conceivably be the true stem group of amphibians. They far surpassed the rhipidistians in longevity, for they survived the extinction that killed off the last of the rhipidistians, and most other species as well, at the end of the Permian Period. It has been estimated that perhaps as many as 95 percent of all species on earth went extinct at this time. Somehow the coelacanths managed to survive and flourish anew in the Mesozoic Era.

During the Mesozoic Era, the Age of Dinosaurs, the coelacanths became largely confined to marine waters, and there, despite their ungainly appearance, they competed successfully with the more streamlined bony fish of the time. But the coelacanths eventually dwindled in numbers as well. The last known genus, named *Macropa*, is found in chalk of late Cretaceous age, some 100 million years ago, in England. Then the coelacanths seemingly vanished from the earth, for we have found no trace of them in the fossil record since that time. They vanished from the fossil record, but not necessarily from among the living. *Macropa*, only ten inches from head to tail, is long extinct. But before dying out, it must have given rise to other species of coelacanths, species that surely can be called living fossils.

Coelacanth Alive

I spent many years studying the externally shelled cephalopod named *Nautilus*, thought by many people to be a living fossil. Nautiluses live in the faraway Western Pacific, off the coral-reefed islands and microcontinents that dot that sun-drenched region. During one of my summers there I worked with two young New Caledonian fishermen who had refurbished a 45-foot boat for deep-water fishing. These enterprising souls had decided that deep-water crabs and fish would afford them an excellent living, for deep-water fish never carry Sigatura fish toxin, a scourge of warm-water fisheries the world over. They outfitted their ship in a professional manner and built deep-water traps. With long lines and bottom-sitting cages, the two fishermen began to fish the steep fore-reef slopes of the New Caledonian Great Barrier Reef. They would drop their meter-cube contraptions of iron and wire to depths of 1000, 1500, even 2000 feet, and retrieve them several days later.

I accompanied these fishermen on several trips and began to live their life: unbelievably hard work in the hot tropical sun and a huge appetite at the end of the day; nights spent rocking at anchor in the trade winds, amid the unrelenting smell of diesel oil, rotting fish, and the never-ending scuttling of shipmate rats and cockroaches. The rewards were many, however. Perhaps the most exciting moment of the day came when the traps were raised. As the cages were slowly brought to the surface, we would strain to see the shapes emerging from the bottom of the sea. I was always amazed at the extraordinary inhabitants of the bottom so far below us. Three creatures typically emerged from the very deep sets: the nautiluses, orange and red of shell, with their many tentacles writhing as they broke into air for the first time; large isopods, identical to the pill bugs so common in our gardens and basements, but much, much bigger (these large crustaceans approached a foot in length); and bottom-dwelling, deep-water sharks, small and with curious tails. The upper lobe, prolonged by the vertebral column, was larger than the lower one, as in all sharks, but still these tails were unlike those of either the familiar swimming sharks or the bony fish of shallower waters. These swarthy sharks, about a foot long, appeared to me to resemble no living fish; they looked instead very much like the long-extinct species of the Devonian Period. In fact, that was the impression I got every time I saw our catches: the deep, fore-reef slopes of these Pacific isles seemed to harbor prehistoric relicts that looked very much like the denizens of the sea some 400 million years ago, when it was dominated by three types of animals: nautiloids that looked very much like the living nautiluses I studied; the long-extinct trilobites, virtually identical in appearance to the large isopods we were catching; and the earliest fishes, consisting of sharks almost identical in appearance to the ones writhing in our cages and some true bony fish. Although the actual creatures had changed, the structural similarities between the extinct Devonian inhabitants and the modern-day creatures of the deep coral reef are striking. This place is a museum of past life. Perhaps more than any other place on earth it is Arthur Conan Doyle's Lost World, where shapes and species of life elsewhere long extinct still flourish.

Fishing is a hard life. The only romance is what you bring to it, and that fades soon. I look back on that period in my life and see it for what it was—unrelenting hard work. But I learned a lot. As we journeyed out from our anchorage I would observe the bottom profile on the echo finder and marvel at how quickly the bottom drops away. The outsides of the coral reefs are among the most precipitous gradients on earth, sheer rocky walls alternating with terraces of rubble and sand, dropping down to depths of over a thousand feet. Then the gradient lessens and the rocky reef scarp gives way to fields of sand and finally mud, dropping endlessly to the abyss.

The reef fronts in these thousand-foot depths are generally bare rock, scoured by strong underwater currents. The surface waters are heated by the tropical sun to bathtub temperatures, but the deep waters are dark and cold. Corals can live no deeper than about 150 feet, so the communities below this critical depth have more in common with the faunas of the abyss than with those of the living reefs.

We lost great quantities of gear on these deep, rocky slopes. As I watched the rapidly dropping contours of the bottom on the echo finder, I often wondered what it would be like to dive on this inaccessible bottom. Far below the limit of scuba gear, the thousand-foot slopes can be visited only by submarine, and few research submarines have ventured into the far Pacific. The edges of these slopes are made up of once-living coral, now pushed downward by the rapidly growing reef above. The thousand-foot contour is a grave of creatures that once lived in the shallows. It is also the home of living creatures very poorly known to us.

There was one last aspect of the thousand-foot bottom that intrigued me in those now long-ago days. Our echo finder gave a picture not only of the depth of the bottom but of the nature of the substrate as well. The ultrasound wave emanating from the transducer on our boat's bottom was reflected back completely by the rocky bottom but only partially by sand bottoms, and very incompletely by muddy bottoms. It was clear that the thousand-foot bottoms of the reefs we fished were made up of bare rock; they had no cover of sediment at all. As a geologist, I knew the implications of this discovery: this place could never have preserved any fossil record. For a marine creature to become a fossil, it must fall into some kind of sedimentary deposit soon after death and become buried there before the remains are destroyed by the myriad scavengers that inhabit the sea. The chance that any creature will become a fossil is small in any circumstance, even if its body falls into soft mud. In an area where sediment is rare or absent, the chance is nil. The rocky scarps of the thousand-foot reef contour are just such places. They are the ideal habitat for a fugitive species—or a living fossil—to hide in.

The coral reefs of the Western Pacific stretch into the Indian Ocean, and the forms and faunas in the two oceans are remarkably similar. In the Indian Ocean, as in the Pacific, some hardy men venture outside the reefs to fish the deep waters. And sometimes they come home with very strange catches indeed.

Such was the case in 1938, off the coast of South Africa. On a hot midsummer's day the trawler *Nerine* pulled into the small port of East London, in the Union of South Africa, with a load of fish trawled up from the offshore deeps. The curator of the New London Museum, Miss M. Courtenay-Latimer, had asked the captain of the *Nerine* to be on the lookout for any strange fish, and to call her if something unusual came up from the depths. On this day the captain had caught a strange and very evil-tempered fish. So he telephoned the museum from dockside as soon as he reached port. Then, as Christmas was fast approaching, he and all of his men save an old Scot left the ship for their various entertainments. Miss Courtenay-Latimer went to the ship, reviewed the pile of fish awaiting her, and was about to leave when she spied a most curious sight. This discovery is best described in her own words:

> I went onto the deck of the trawler *Nerine* and there I found a pile of small sharks, spiny dogfish, rays, starfish, and rat tail fish. I said to the old gentleman, "They all look much the same, perhaps I won't bother with these today"; then, as I moved them, I saw a blue fin and pushing off the fish, the most beautiful

> fish I had ever seen was revealed. It was 5 feet long and a pale mauvy blue with iridescent silver markings. "What is this?" I asked the old gentlemen. "Well, lass," said he, "this fish snapped at the Captain's fingers as he looked at it in the trawler net. It was trawled with a ton and a half of fish plus all these dogfish and others." "Oh," I said, "this I will definitely take to the museum and I shall not worry with the rest." I called Enoch, my native boy, and with a bag we had brought down he and I placed it in the bag and carried it to the taxi. Here to my amazement, the taxi man said, "No stinking fish in my taxi!" I said, "Well you can go, the fish is not stinking. I will call another taxi." With that, he allowed us to put it into the boot of the taxi.[1]

Since the fish weighed 127 pounds, one can imagine the interesting scene as this young curator and her assistant bagged and transported it from boat to taxi.

And so off into history rode Miss M. Courtenay-Latimer and the big blue fish. Unfortunately, the curator soon found that her big blue fish was rapidly becoming a big gray fish in the heat of the day. Courtenay-Latimer now had a very pressing problem indeed. She had an enormous fish of unknown affinity that had to be preserved somehow, and fast.

When she reached the museum, Miss Courtenay-Latimer got the fish hoisted onto a table and then went immediately to her reference books, for she knew that this fish was something extraordinary. The fins were most unusual; they looked almost like arms, so that the fish had a lizardlike appearance. The fish's scales were also unusual—large and heavily ossified, covered with toothlike protuberances—and the head was encased in bony armor. The most similar South African fish she could think of was the lungfish, but no known lungfish had the appearance of the rapidly decomposing individual sprawled on her table. In haste Courtenay-Latimer measured and sketched the fish, and then, lacking a freezer, decided to take it a taxidermist. When one has a decomposing fish on a hot day, the simplest course of action is to discard it, and that option surely must have tempted the young curator. There would be no story to tell if Miss Courtenay-Latimer had been less determined to save a strange specimen that she knew to be important.

Courtenay-Latimer and the taxidermist placed cloth strips soaked in formalin about the fish, hoping to preserve it until expert ichthyologists could make a more thorough examination.

[1]Reprinted with permission from M. Courtenay-Latimer, *My Story of the First Coelacanth,* Occasional Papers of the California Academy of Sciences, no. 134 (San Francisco, 1979).

She then wrote a letter to J. L. B. Smith, a chemist-turned-ichthyologist who had collected fish for the East London Museum, and who was at that time publishing taxonomic monographs on the South African fish fauna. Courtenay-Latimer included the sketches in her letter and then settled down to wait. Unfortunately, the bacteria multiplying within the tissues of her fish would not wait. The thick scales inhibited the formalin from impregnating the interior tissues, so that within three days of its capture the large fish was in an advanced stage of decomposition. By December 26, four days after its capture, Courtenay-Latimer decided that only the outer skin and head were worth saving. The rest of the fish, including gills and internal organs, was put into the trash to be dumped into the sea.

Courtenay-Latimer's letter and sketches describing her strange acquisition did not reach Smith until January 3. James

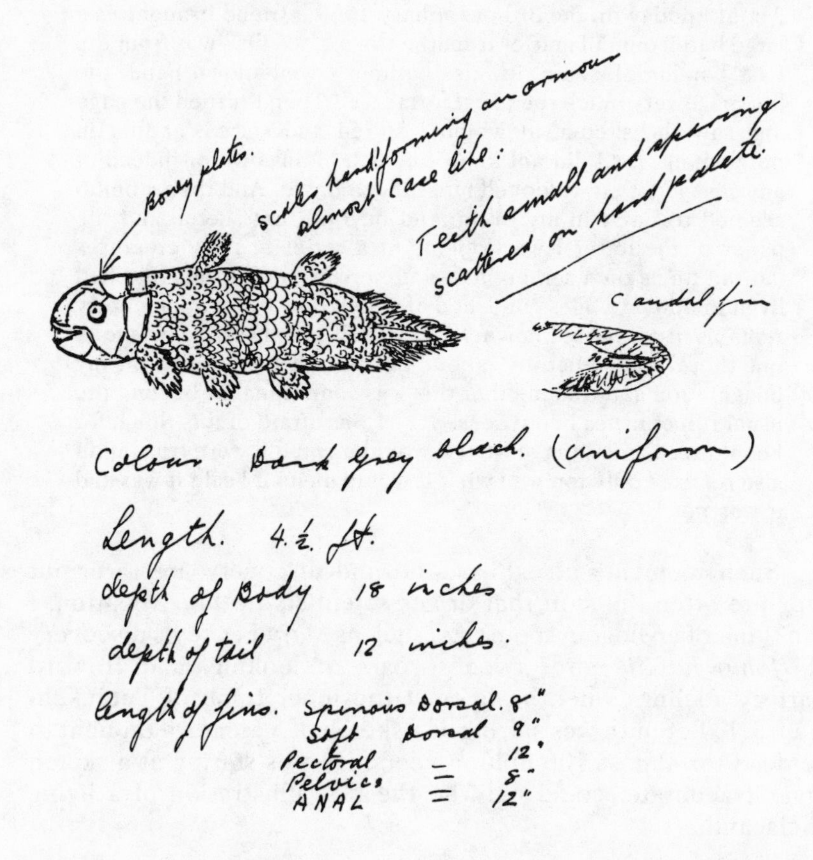

Courtenay-Latimer's sketch of the first coelacanth in her letter to Smith. (From P. H. Greenwood, A Living Fossil Fish: The Coelacanth *(London: British Museum [National History, 1988].)*

Leonard Briefly Smith was an exceptional man. He has been described in many ways with varying degrees of admiration in the years since this first capture of a living coelacanth. Michael Lagios and John McCosker, ichthyologists and members of a recent expedition to capture living coelacanths, described Smith as "adventurer and ichthyologist," expressing an opinion that came to be widely shared in the scientific community as events unfolded. Whatever else can be said about him, he was the right man at the right time: not only did he know modern fishes, but he had more than passing knowledge of many types of long-extinct fish as well. And upon receiving Courtenay-Latimer's sketches he immediately knew that he was dealing with a hoary creature indeed.

In his somewhat hyperbolic account of those heady days, Smith described his reaction when he read Courtenay-Latimer's letter:

> About midday on the 3rd of January 1939 a friend brought us a large batch of mail matter from the town. . . . One was from the East London Museum, in Miss Latimer's well-known hand, the first page very much the usual form. . . . Then I turned the page and saw the sketch, at which I stared and stared, at first in puzzlement, for I did not know any fish of our own or indeed of any seas like that; it looked more like a lizard. And then a bomb seemed to burst in my brain, and beyond that sketch and the paper of the letter I was looking at a series of fishy creatures flashed up as on a screen, fishes no longer here, fishes that had lived in dim past ages gone, and of which only often fragmentary remains in rocks are known. I told myself sternly not to be a fool, but there was something about the sketch that seized on my imagination and told me that this was something far beyond the usual run of fishes in our seas. . . . I was afraid of this thing, for I could see something of what it would mean if it were true, and I also realized only too well what it would mean if I said it was and it was not.[2]

Such moments of exhilaration and discovery are rare, but they are often the stuff that drives scientists to their vocation. I can think of analogous moments, such as Mary Leakey's discovery of *Homo habilis* after twenty years of looking, and Howard Carter's feelings when he opened the inner tomb of Tutankahmen. J. L. B. Smith was staring at a sketch of a creature thought to be dead for almost 100 million years. He was staring at a sketch that, if accurate, could only be the first illustration of a living coelacanth.

[2]Reprinted with permission from J. L. B. Smith, *The Search beneath the Sea* (New York: Henry Holt, 1956); p. 62.

Smith immediately telegraphed to the East London Museum: "MOST IMPORTANT PRESERVE SKELETON AND GILLS FISH DESCRIBED." Unfortunately, by this time the bones and gills were long gone into the sea. Only the skin remained.

In early January, Smith made the first announcement of his identification of the great fish as a still-living coelacanth in letters to both Courtenay-Latimer and K. Barnard, assistant director of the South African Museum. Barnard, also an ichthyologist, was skeptical, to say the least; how could a prehistoric five-foot-long fish have evaded not only the nets of all commercial fishermen but 100 million years of the fossil record? Such reactions were apparently common in the early days of Smith's heady discovery. It took enormous intellectual courage to continue to insist that the fish was a coelacanth, in the face of almost universal ridicule. Perhaps, out of fear of what he might find, Smith put off visiting the New London Museum to confirm or deny the identification he had made, at this time still based only on a single sketch sent by Courtenay-Latimer. It was not until February 1939 that Smith actually saw the now-stuffed fish. He described his first view of it:

> We went straight to the Museum. Miss Latimer was out for the moment, the caretaker ushered us into the inner room and there

Preserved coelacanth. (Courtesy Department of Library Services, American Museum of Natural History.)

> was the—Coelacanth, yes, God! Although I had come prepared, that first sight hit me like a white-hot blast and made me feel shaky and queer, my body tingled. I stood as if stricken to stone. Yes, there was not a shadow of doubt, scale by scale, bone by bone, fin by fin, it was a true Coelacanth. It could have been one of those creatures of 200 million years ago come alive again. I forgot everything else and just looked and looked, and then almost fearfully went close up and touched and stroked, while my wife watched in silence. Miss Latimer came in and greeted us warmly. It was only then that speech came back, the exact words I have forgotten, but it was to tell them that it was true, it was really true, it was unquestionably a Coelacanth. Not even I could doubt any more.[3]

Smith's problem now was to describe and publicize the discovery. He was under several conflicting pressures. On the one hand, he had taken the risk of making the controversial identification, so (he considered) it was his prerogative to name and describe the great fish. Therefore he wanted no photos of the fish taken and no popular account to reach the newspapers until he had formally described the coelacanth in a scientific journal. On the other hand, the board of the New London Museum, after initial skepticism, wanted as much publicity as possible to promote the museum. Smith and Courtenay-Latimer thus arrived at a compromise decision: they let a single reporter describe and photograph the fish, with the understanding that the description would appear in the local paper only, and that the photo would not be reprinted. Imagine, then, Smith's joy on learning that the photo had been promptly sold to any and all comers throughout the world. "MISSING LINK FOUND," bellowed headlines throughout the world.

Having been thus burned in this first skirmish with the popular press, Smith decided to finish the painstaking task of minute observation and description of the fish's anatomy away from the glare of publicity, in the privacy of his home. Perhaps it was this decision that alienated so many ichthyologists, for as news spread of the wondrous return to life of this relict of the Paleozoic seas, many ichthyologists looked forward to collaborating in the study of the great find. But Smith insisted on working up all aspects of the fish by himself. The fish was sent by rail to Smith's home in Grahamstown, South Africa, under police guard—a testament to the power of the press, for the coelacanth had become a cause célèbre in South Africa, and surely an antidote to the news of war clouds gathering over Europe. But Smith was able to keep the fish in his possession only about two months. Public pressure to view

[3] Ibid., p. 73.

the coelacanth was so great that in May he was forced to return it to the New London Museum, which put it on public display.

The first scientific description of the specimen was published by the highly respected scientific journal *Nature* on March 18, 1939. Accompanying the brief article, headed "A Living Fish of Mesozoic Type," was a large picture of the fish. Smith named the fish *Latimeria chalumnae*, new genus and new species. All scientific doubts (of which there had been many) were apparently silenced with the publication of this article. But many questions remained, not the least of which concerned the way this living creature had remained hidden so long from zoologists as well as paleontologists. And if one specimen had demonstrably been alive a few months ago, where were the rest of the living members of this species?

The specimen itself was a very mixed blessing. Without the interior parts and the gills, Smith lacked several important clues that would have shed much light on the evolution of this stock of fishes. If these questions were ever to be resolved, it was imperative that a second specimen be obtained. But where?

The Search for the Second Coelacanth

M. Courtenay-Latimer and J. L. B. Smith had ascertained that the original coelacanth had been trawled at 40 meters near the South African coast. Thus it had come from a heavily fished region. It seemed highly unlikely that any significant stock of such large and peculiar fish could have escaped the notice of fishermen, unless the coelacanth population was vanishingly small. The other possibility was that this particular specimen was caught far from its real habitat. This seemed the most likely explanation to Smith.

The discovery of the coelacanth, coming as it did a few months before the outbreak of World War II, was soon overshadowed by the tragic news from Europe and then from the Pacific. Smith, determined to find more specimens, spent the war years talking with fishermen and naturalists—anyone acquainted with the fish fauna of the South African coast. Although he heard rumors of large, strange fish that had been caught along the coast from time to time, he never found any hard evidence. As the years drifted by, frustration only fueled Smith's obsession.

When the war ended, Smith was commissioned to write a book about South and East African marine fishes. He welcomed this task as an opportunity to continue his search for the home territory of the coelacanth. To encourage cooperation, Smith authorized a reward of £100 for any coelacanth captured, and had numerous posters and leaflets printed, each with a photo of the fish, and distributed them throughout the coastal regions of South and East Africa. But as the years rolled on and no more coelacanths surfaced, it became apparent to Smith that the African coastal waters were not the home of this elusive fish; heavy fishing efforts in the region that had yielded the first African coelacanth never produced a second. Some fisheries experts suggested that the coelacanth was a member of the deep-sea fauna, living perhaps at depths of 1000 meters or more, and that the specimen caught in 1938 was a stray. According to this argument, coelacanths did indeed live off the African coast, but at depths beyond the limits of commercial fisheries operations. Smith scoffed. The characteristics of the fins and the presence of bony head armor, he argued, indicated that the coelacanth was more closely allied to the shallow reef biota than to the deep sea.

But if the coelacanth did not live along the African coast, where did it live? Smith's only clue came from ocean circulation patterns. He noted that strong surface currents often brought tropical species southward from the more equatorial regions of East Africa. Why couldn't these currents have brought the 1938 specimen south as well? Smith concluded that the home of the

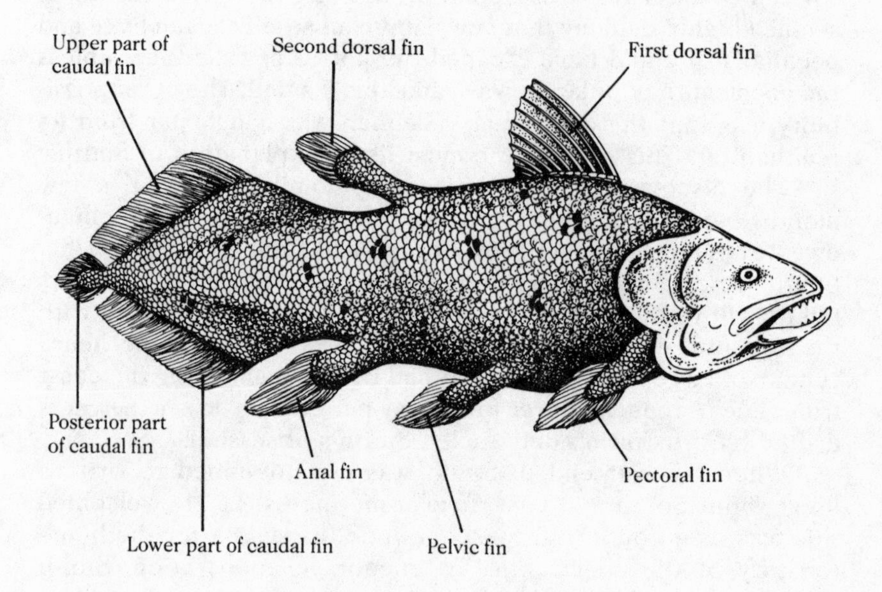

Anatomy of Latimeria, *the coelacanth.*

PREMIO £ 100 REWARD
RÉCOMPENSE

Examine este peixe com cuidado. Talvez lhe dê sorte. Repare nos dois rabos que possui e nas suas estranhas barbatanas. O único exemplar que a ciência encontrou tinha, de comprimento, 160 centímetros. Mas já houve quem visse outros. Se tiver a sorte de apanhar ou encontrar algum NÃO O CORTE NEM O LIMPE DE QUALQUER MODO — conduza-o imediatamente, inteiro, a um frigorífico ou peça a pessoa competente que dele se ocupe. Solicite, ao mesmo tempo, a essa pessoa, que avise imediatamente, por meio de telgrama, o professor J. L. B. Smith, da Rhodes University, Grahamstown, União Sul-Africana.

Os dois primeiros especimes serão pagos à razão de 10.000$, cada, sendo o pagamento garantido pela Rhodes University e pelo South African Council for Scientific and Industrial Research. Se conseguir obter mais de dois, conserve-os todos, visto terem grande valor, para fins científicos, e as suas canseiras serão bem recompensadas.

COELACANTH

Look carefully at this fish. It may bring you good fortune. Note th[illegible]uliar double tail, and the fins. The only one ever saved for science was 5 ft (160 cm.) long. Others [illegible] been seen. If you have the good fortune to catch or find one DO NOT CUT OR CLEAN IT ANY WA[illegible] get it whole at once to a cold storage or to some responsible official who can care for it, and ask [illegible]o notify Professor J. L. B. Smith of Rhodes University Grahamstown, Union of S. A., immediately by tel[illegible]aph. For the first 2 specimens £ 100 (10.000 Esc.) each will be paid, guaranteed by Rhodes University a[illegible] by the South African Council for Scientific and Industrial Research. If you get more than 2, sav[illegible] the[illegible]ail, as every one is valuable for scientific purposes and you will be well paid.

Veuillez remarquer avec attention ce poisson. Il pourra vous apporter bonne chance, peut être. Regardez les deux queuex qu'il possède et ses étranges nageoires. Le seul exemplaire que la science a trouvé avait, de longueur, 160 centimètres. Cependant d'autres ont trouvés quelques exemplaires en plus.

Si jamais vous avez la chance d'en trouver un NE LE DÉCOUPEZ PAS NI NE LE NETTOYEZ D'AUCUNE FAÇON, conduisez-le immediatement, tout entier, a un frigorifique ou glacière en demandant a une personne competante de s'en occuper. Simultanement veuillez prier a cette personne de faire part telegraphiquement à Mr. le Professeus J. L. B. Smith, de la Rhodes University, Grahamstown, Union Sud-Africaine.

Le deux premiers exemplaires seront payés à la raison de £ 100 chaque dont le payment est garanti par la Rhodes University et par le South African Council for Scientific and Industrial Research.

Si, jamais il vous est possible d'en obtenir plus de deux, nous vous serions très grés de les conserver vu qu'ils sont d'une très grande valeur pour fins scientifiques, et, neanmoins les fatigues pour obtantion seront bien recompensées.

This reward poster, circulated along the African coast, resulted in the capture of the second coelacanth to be found alive. (From Margaret M. Smith, The influence of the coelacanth on African ichthyology, Occasional Papers of the California Academy of Sciences, *No. 134 [Dec. 22, 1979], p. 13.)*

coelacanths lay north and east of the first capture site: either near Madagascar or off some of the smaller islands in the Indian Ocean. One group of islands, poorly known, most piqued his interest: the Comoro Islands.

In the early 1950s Smith met Eric Hunt, a trader who worked the Comoros, and gave him a batch of the reward leaflets to distribute among the fishermen there. Hunt did so, and the response was almost immediate. On Christmas Eve, 1952, J. L. B. Smith received the following telegram: "HAVE FIVE FOOT SPECIMEN COELACANTH INJECTED FORMALIN HERE KILLED 20TH ADVISE REPLY, HUNT, DZAOUDI."

At the time he received the message, Smith was aboard a fishing vessel off South Africa. Neither Smith nor the fishermen with him had ever heard of a place called Dzaoudi. A frantic search of marine charts finally yielded the location: sure enough, the Comoro Islands.

Smith knew that the Comoros were a string of small islands about 12 degrees south of the equator, between East Africa and Madagascar. He knew as well that these islands were extraordinarily hot, rocky patches of real estate, far from land in the Indian Ocean, and that living conditions and facilities were primitive. Not only was there no refrigeration available, but even the fixatives necessary to preserve the great fish would be scarce, if they were available at all. Although Hunt's telegram said that he had already injected his specimen with formalin, Smith had no way of knowing the actual state of preservation of the fish, or even if it was indeed a coelacanth, for all Hunt had to go on was the photo on the reward leaflet. Reporters soon got wind of the possibility of a second coelacanth catch, fourteen years after the first, and Smith was inundated by a new generation of journalists seeking a scoop about the "missing link."

Smith was in a quandary. He was several thousand miles away from this specimen, which was now residing, if it existed at all, in the custody of a foreign government. Time was clearly not on his side. He had to travel immediately to the Comoros, but he had no way of getting there. So there he sat, on Christmas Eve, facing a most frustrating problem.

Eric Hunt, waiting in the Comoros, was having difficulties of his own. He had obtained the fish only by a most unlikely string of circumstances. Hunt had distributed the coelacanth leaflets to various fishermen and marketplaces. Comoro Island fishermen were astounded by the reward, for to any of these subsistence-level people, £100 was greater than any king's ransom, more by far than they could hope to earn by many years of toil at sea. And perhaps they were astounded as well by the identity of the fish being sought, for generations of Comoro Island fishermen knew this fish. They called it *kombessa*, and it was known for its terrible smell, the oiliness of its flesh, and the great difficulty of killing it. It was rare—only one or two were caught each year—but not a fish one was likely to forget.

On the moonless night of December 20, 1952, a fisherman named Ahmed Hussein and a friend put their small boat out from shore and rowed out to the fishing grounds. The Comoros are volcanic islands that rise abruptly from the ocean bottom. Because the sides of the islands are very steep, extremely deep water is very close to land, and a good thing: survival for these fishermen often depended on a rapid return to land in the face of a sudden squall. The winds at these latitudes are very treacherous, particularly during the typhoon season from December through March. This night the two fishermen baited their hooks and dropped the coral-weighted lines into the blackness of the water. They were soon rewarded by an enormous tug from far below, and found themselves in a great fight as a large fish struggled against capture. The fishermen must have been astounded by the size of the fish that they finally dragged to the surface. They took no chances with the snapping brute, armed as it was with a terrible temper backed up by huge, needlelike teeth: they bashed its brains in with an oar before attempting to haul it into the small boat. This in itself was no small feat, for the head of the fish was heavily armored with thick bony plating. Still the fish clung to life, as if refusing to die. When finally they had dragged it into their boat and rowed back to their village, the went off to bed without bothering to clean the fish. And so this second capture of the rarest of fishes, object of fourteen years' search, sat outside a small hut through the night, slowing decaying. But it was still intact (if somewhat battered about the head), for the fishermen's failure to clean the fish immediately saved its internal organs.

In the morning Hussein took his huge fish to the marketplace to sell. He was about to cut it up when another fisherman pointed out that the fish looked very much like the picture on the leaflet being handed out by Captain Hunt. The instructions on the leaflet were explicit: Do not cut or clean the fish, but take it to some responsible person. The fishermen knew that at that moment Captain Hunt was anchored on the other side of the island, so Hussein and his friend decided to hike the fish over the island and present it to him.

One can only imagine the nightmare trek that ensued. The fish weighed only slightly less than 100 pounds. The fishermen dragged it more than twenty-five miles in searing heat. The sun was setting by the time they reached Hunt's ship.

The fish was stinking and the fishermen wanted their money. But Hunt immediately recognized that it was a coelacanth, though it differed somewhat in appearance from the 1938 specimen: this new catch had only one dorsal fin. Hunt immediately had his crew cut and salt the fish in an attempt to slow the putrification. As he was docked on an island and had no other preservative and no communication facilities, Hunt lifted anchor and set sail in his schooner for the capitol of the Comoro Islands. When he arrived he informed the authorities of the catch, and obtained formalin from a local doctor to inject the fish. He then sent his cable to Smith in South Africa.

Hunt was walking a very narrow tightrope indeed. On the one hand, he depended on the goodwill of the local administration for his livelihood; on the other hand, he had a priceless scientific specimen that he was determined to turn over to the one man who most wanted to see it: Smith. The Comoros at that time were a colony of France, so Hunt had to deal with somewhat skeptical French officials while he awaited a response—any response—from Smith. Hunt convinced the governor of the Comoros that Smith would soon be arriving by plane to take the fish back to South Africa. One of the fascinating aspects of this story is the great trust between Smith and Hunt, who up to that time were only casual acquaintances. Smith gambled his scientific and political credibility on Hunt's ability to recognize a coelacanth when he saw one. Hunt gambled his credibility with the local government on the assumption that Smith would somehow commandeer an airplane and pick up the fish.

Smith was doing his best to organize one. There were no scheduled flights to the Comoros, and Smith had nowhere near enough money to charter a plane. He finally decided to go to the top: he appealed to the prime minister of the Union of South Africa, Daniel F. Malan, who, surprisingly, agreed to provide a government plane. And so early on December 29, 1952, Smith roared off in a Dakota military transport, the sole passenger on a trip of many hours, to bring back a dead fish.

Upon arrival in the Comoros, Smith had to endure a diplomatic welcome before he could see the fish and determine if it really was a coelacanth. Smith described his reaction when he finally arrived at Hunt's schooner:

> Hunt pointed to a large coffin-like box near the mast, and I knew it must be in there. They picked up the box and put it on the hatchcover, just in front of me, a foot above the deck, and Hunt pulled away the lid. I saw a sea of cotton-wool, the fish was covered by it. My whole life welled up in a terrible flood of fear and agony, and I could not speak or move. They all stood staring at me, but I could not bring myself to touch it; and, after standing as if stricken, motioned them to open it, when Hunt and a

sailor jumped as if electrified and peeled away that enveloping white shroud. God, yes! It was true! I saw first the unmistakable tubercles and the large scales, then the bones of the head, the spiny fins. It was true![4]

With alacrity the fish was repacked and hustled onto the waiting plane, for Smith knew it would soon dawn on the French that any fish valuable enough to be sent for by military transport might be a fish worth keeping. Smith arrived back in South Africa late the same day with his treasure.

Smith's more leisured inspection of the fish suggested that this coelacanth, with its single dorsal fin, represented an entirely different species, and even a new genus. Smith named this new coelacanth *Malania anjouanae*, in honor of the prime minister of South Africa. The prime minister received a special viewing of the

[4]Ibid.

J. L. B. Smith (kneeling, center) and Eric Hunt (at extreme left) with the second coelacanth. (From Margaret M. Smith, The influence of the coelacanth on African ichthyology, Occasional Papers of the California Academy of Sciences, *No. 134 [Dec. 22, 1979], p. 15.)*

salted, formalin-injected fish in his backyard, courtesy of the triumphant Smith. The Prime Minister's joy at being memorialized and having his name forever linked with a vicious, nearly extinct fish with a uniquely disgusting smell has not been recorded, but it may have been with some relief that the eminent politician soon learned that other ichthyologists did not accept Smith's assessment of the second coelacanth as a distinct genus and species. The absence of the first dorsal fin in the new specimen, by which Smith justified his new name, was determined to have been caused by an accident early in the fish's life. *Malania anjouanae* Smith became an invalid synonym of *Latimeria chalumnae* Smith. Only a single coelacanth species was still alive.

The Fate of a Living Fossil

J. L. B. Smith's association with coelacanths ended after the capture of the second specimen. Smith very much wanted to continue his virtual monopoly on coelacanth study, but his somewhat high-handed tactics eventually caught up with him. He had high hopes of capturing more of these giant fishes, and wanted eventually to expand the search operations to other island groups, especially to the nearby Aldabra Islands and Madagascar. But after the capture of the second coelacanth and Smith's flight to collect it, the French press denounced him as a pirate, thief, and worse. Having lost one specimen, the French now viewed coelacanths as national treasures, and banned foreigners from fishing for or acquiring them. His search for more coelacanths thus obstructed and his success turned into an international scandal, Smith went home and had nothing more to do with the mystery of the coelacanth.

The capture of the second coelacanth revealed the true home of these relicts of the past. The first specimen, trawled up from relatively shallow water off South Africa, was clearly a long way from home, for all other coelacanths found since have come from the Comoro Islands.

With Smith no longer involved, the 1952 capture of the second coelacanth set up a piscine gold rush in the Comoros. Once word got out that coelacanths were alive and well and living off the Comoros, many major museums clamored for specimens. Coelacanths became one of the islands' major exports. Between 1952 and 1975, 83 specimens were hauled from the depths off the Comoro Islands, and many fishermen began to specialize in them, for a single specimen provided a year's salary in reward money. With such wealth to be had, the coelacanth fishermen perfected their techniques and the catch rate escalated; between 1975 and about 1985 another 120 were captured.

As a slow but regular procession of coelacanths began to arrive from the Comoros during the 1950s and 1960s, scientists began to arrive at an understanding of the biology of this fish. One of the first things they learned was that Smith was mistaken in believing that the coelacanth was a shallow-water reef fish. The majority of captures have been made at depths between 600 and 1200 feet, and some fish have been found as deep as 2000 feet. The coelacanth is clearly not a shallow-water fish.

A second surprise concerned the reproductive habits of the coelacanth. Coelacanth eggs have turned out to be as large as grapefruits. They are never released from the mother, for, like sharks, the young coelacanths hatch from the eggs internally and then emerge alive from the mother. Such a system reduces juvenile mortality, for the young have already attained a good size when they hatch. A disadvantage of this type of reproductive strategy is that clutch sizes are invariably small and the gestation period is long. Each female coelacanth can produce only a small number of young, and if fishermen become overenthusiastic, they can easily deplete the population.

One of the most interesting and long-awaited observations concerned the way the coelacanths use their lobed fins to move through the water. If the coelacanth did indeed represent the transition from fish to amphibians, much useful information was to be gleaned from the morphology, the musculature, and especially the mobility and strength of the pectoral and pelvic fins. This information, however, could come only from observations of living specimens, and all of the coelacanths fished up from the deeps died soon after capture. It was many years before ichthyologists were able to view a living coelacanth. Eventually several of the large fish were kept alive for some hours—not many, but enough to permit naturalists to view the last moments of life of these large fish. At last they could see how the coelacanths used their large, fleshy fins to swim. The fish used the lobed fins on their sides to fend off objects, much as we use our arms. But the captured specimens never used their fins for walking on the bottom.

A succession of research expeditions to the Indian Ocean provided more tantalizing clues. One striking finding was that the Comoros sustain an extremely low population of fishes in comparison with other islands of similar size in the Indian Ocean. And not only are there fewer fish in the Comoros, but the percentage of larger fish predators is also very low. The strange makeup of the fish fauna in the Comoros may be related to the nature of the seawater in the area. As the Comoro Islands are volcanic, they have a large number of underwater caves. A great deal of groundwater seems to seep out of the sides of the island and into these caves. The seawater immediately surrounding the Comoros can be slightly brackish when rainfall is heavy, and it has been speculated that the coelacanths live in the underwater caves, bathed in a dilute mixture of seawater and rainwater. Few modern fishes can tolerate such a mix.

It thus appears that coelacanths have survived in an environment where there is little competition with more advanced fishes and little predation by other fishes as well. Charles Darwin would have been gratified to hear of these findings, for he considered that living fossils remained uniform in their morphology over long periods of time precisely because they were adapted to just these types of environments, where competition with other species was low or absent. The coelacanth lives virtually alone in the barren, thousand-foot depths off the Comoro Islands, a stark, lifeless place more akin to the surface of the moon than the sea bottom of our planet.

One of J. L. B. Smith's most fervent hopes was to see a living coelacanth. Alas, this was was never granted, and until recently no scientist saw any but dying specimens, for no coelacanth lived even a day in captivity. It would take a submarine actually to see a living coelacanth in its habitat. Happily enough, in 1987 just such an expedition was launched.

Diving in a tiny submersible, the German naturalist Hans Fricke succeeded in observing and photographing six living coelacanths during dives off Grand Comoro Island. He followed coelacanths for as long as six hours, observing and filming their odd movements as they swam forward, backward, even upside down. He saw no coelacanth crawl on the bottom with its lobed fins, as Smith had supposed it might. An unexpected highlight was the finding that the coelacanth seems to produce an electric field around its body, as the electric eel does. Fricke also came to realize how rare these great fish are, for he had to make twenty-two dives before he saw one.

The year 1988 marked the fiftieth anniversary of the first capture of a living coelacanth. In a thoughtful article in *Nature*, Peter Forey, a paleontologist with the British Museum, summarized the current state of our knowledge about both living and extinct coelacanths. The great excitement about coelacanths came from the assumption that they were "missing links" between the first amphibians and the fish that spawned them. But our knowledge of the fossil record is much better now than it was fifty years ago, and our understanding of the coelacanth's anatomy has vastly improved as well. We now know that *Latimeria*, the living coelacanth, is substantially different from what we suppose the immediate ancestor of the first amphibians looked like. Some evidence even suggests that the living coelacanth is more closely related to the cartilaginous fishes—the sharks, rays, and skates—than to the bony fishes, and hence is far from the evolutionary line from which the first amphibians (and ultimately ourselves) emerged. Other scientists argue that the living coelacanth is related to the direct ancestors of the amphibians, but has greatly evolved in the intervening 400 million years. No one, however, doubts that *Latimeria* can continue to teach us much about the past—if it can be saved from extinction. We have absolutely no inkling of the size of the *Latimeria* population, or of its ability to withstand the current fishing efforts of the Comoro islanders. The monetary value of the coelacanth, which has increased in proportion to its scientific interest, is probably a far greater threat than anything in its natural environment. What tragic irony it will be if the great blue fish is exterminated from the earth, after hundreds of millions of years, not by natural forces in the thousand-foot-deep environment but by the very scientists who wish to understand its longevity.

Envoi

A year and a season have passed since my late-May encounter with the Cretaceous-Tertiary boundary at Château Bellecq. It's September 1990 and I'm once again in France, after weeks of collecting fossils on the beaches of Zumaya, Hendaye, and Biarritz. The fossils are in my sample bags now, along with the first draft of this book. These golden September days, before I tackle the task of tidying the manuscript, are a good time to reflect on all that has taken place during the year just past, and about the themes of survival and death that I have lived with while I have chronicaled the Methuselahs. Why do we still have horseshoe crabs and magnolias, brachiopods and oysters, nautiluses but not ammonites? There is no single reason. Some species found refuges, others evolved shapes and forms that have been continually successful amid the sweeping changes of the last half-billion years on earth. One word sums up the reason for many of the survivals: luck.

I sit back on my bench, watching another living fossil. A dragonfly flits by and finally alights on the ground near me. Its kind were common in the Carboniferous coal swamps, and they are common still. Did they survive by savagery and skill—or by luck? The dragonfly flies away in alarm as a human horde descends on my musings.

My park bench becomes an island amid a swirling sea of unbridled energy, as gangs of French schoolchildren run screaming sprints beneath the yellowing leaves of the huge sycamore trees overhead, exhorted by whistle-toting teachers from the nearby *lycée.* I look up from the children to gaze once again at the imposing Hall of Zoology, part of the immense complex that makes up the Museums of Natural History in the Jardin des Plantes in Paris. My bench faces the length of the zoology building, and from the yellow limestone facade a row of statues stare stonily down at the children running, the homeless man sleeping, and the lone American musing on the green park bench. One face especially draws my attention: on the far right side of the building the stone features of the nineteenth-century geologist and zoologist Alcide d'Orbigny gazes out across the magnificent gardens, across the blooming dahlias toward the distant Hall of Paleontology, barely visible through the trees at the far end of the park. I'm not sure why d'Orbigny was put here, on the walls of the zoology building, rather than on the distant paleontology building. I suspect he would rather be with his collections.

This immense park was set aside by the power of the French Empire centuries ago; it is now a monument to the awakening of French science, which began after the Revolution, when great men marshaled ideas that still greatly affect my science, as well as many other branches of learning. It was here that Lamarck formulated one of the first theories about evolution; that Buffon built the giant zoological collections that vastly increased the understanding of nature; that Cuvier started the science of comparative anatomy and proposed the first theory to account for mass extinctions; and that the great pioneering paleontologist Alcide d'Orbigny studied the succession of fossils from the Mesozoic strata of Europe in an effort to create a table of strata recognizable throughout the world.

D'Orbigny was one of the founders of biostratigraphy, the science that dates the succession of sedimentary rocks through the observation of fossils. He cataloged and described the fossils from France and Germany, and in the process arrived at a system for subdividing strata which is still in use today. These subdivisions, which he called stages, are the most refined units of geological time recognizable throughout the world. He identified these units with the unwavering confidence that they could be used everywhere on earth as page markers in the great book of time and rock, the stratigraphic record. His concept was simple: God created a series of animals and plants and allowed them to cover the earth and live for a time, and leave behind their fossilized remains in the rock as a record of their time on earth; and then God killed them all in a sudden catastrophe. But God in his wisdom despaired before the now-barren earth, and once again repopulated the seas with swimmers and the land with crawlers and runners, and saw once again a fruitful world. And once again this new fauna lived for a time, and left its fossils, and then was killed off. In this way d'Orbigny explained the table of fossils that he could recognize everywhere in France, and surely in the rest of the world as well. It is a great irony that d'Orbigny's stages, based on reasoning that crumbles in the light of evidence uncovered since his day, are still useful.

D'Orbigny certainly knew about the Cretaceous Period and the great extinction that ended it. More than a century and a half before me he walked the rocky coasts of southern France and collected fossils from the quarries where I have more recently sweated. The expedition from which I have just returned is bittersweet to me, for it marks my last scientific visit to these sites; I have said my farewells to Zumaya, Hendaye, and Biarritz. My work there is finished, or as finished as one gets in my profession, for there are always new questions and new leads that can be followed. But the time has come to turn to new scientific problems, as soon as I finish up the last phases of the decade's research. I am in Paris to visit the museum collections of the great Hall of Paleontology in this park, for now I must publish the results of my long research into the causes and consequences of the extinction at the end of the Cretaceous. I must write detailed descriptions of all the ammonite species I have collected during my research, and I can do that job only by comparing them with other fossil collections and by examining the various taxa and type specimens that serve as the standards and examples of my fossil species. Just as the lives of currently living species are affected not only by the other species that live among them but by the events and species that preceded them as well, my work builds on the work of those scientists who work beside me and came before me.

I was admitted into the paleontological museum early this morning and found it to be a mélange of the old and new; dusty collections in ancient offices sat next to computers. The keeper of the fossil collections, Monsieur Fisher, received me in his office and told me how the collections are arranged. I was then admitted into the collections, and I spent several hours chasing down the various fossil ammonites that I needed to study. Finally, near lunchtime, I was interrupted by Monsieur Fisher with news that perhaps some of the things I needed to see were to be found in another room. Would Monsieur Ward care to follow?

We passed through the areas open to the public, first moving through the giant boneyard of Georges Cuvier, where skeletons of every conceivable family of vertebrates sit jumbled together, proof of his new nineteenth-century science of comparative anatomy; then up a floor, through the Hall of Fossils, walking, too rapidly for me, past immense dinosaur skeletons and fossils of every kind. I had to stop once or twice when I saw old friends: an immense preserved coelacanth was there, surrounded by skeletons of its fossil ancestors. I walked past horseshoe crabs, both ancient and recent, and saw innumerable nautiluses, ammonites, and brachiopods laid out in old-style cabinets. We finally reached an ornate door with stern signs warning the public away. Monsieur Fisher led me into the room, a place of high ceilings and large windows overlooking the blooming gardens. I knew instantly where we were, even as Monsieur Fisher welcomed me to the office and private collections of Alcide d'Orbigny.

Ammonites lay scattered over every possible surface of the large room. The sides of the room were covered with giant cases of them, visible through glass doors. A fine dust covered everything. We searched through the drawers for a particular specimen and finally found the one I needed. D'Orbigny had collected it long before there were cars or airplanes, long before Custer died, long before the American Civil War, before Darwin published his great theory, before the California gold rush. An accompanying slip of paper, yellow with age, was covered with faded handwriting I assumed to be d'Orbigny's. The beautiful script of a scientist long dead sent information down through the generations to the heart and mind of a scientific descendant.

I jotted down the information about this fossil, an ammonite unremarkable, a specimen no different from many I had collected in recent weeks, save that it had been collected over 150 years ago. And not for the first or last time I appreciated not only the immensity of geologic time but its apparent discontinuity. Time seems to run on many tracks at many speeds. To me, an American in Paris in 1990, the fossil in my hand was old beyond belief, for it had sat in this museum for a century and a half. But the creature itself, which died and fell into the muddy bottom of a forgotten seabed—how can any human understand how long ago 66 million years ago is?

A slight noise made me look up, and I was surprised to see that Monsieur Fisher and I were not alone in d'Orbigny's room. In the far corner an old man sat at a desk, studying a box of ammonites. I went over and introduced myself. He told me that he was continuing the curatorial work on the d'Orbigny collections begun more than century before. The collections that d'Orbigny gathered during the epic period of nineteenth-century canal building and stone quarrying were so vast that generations of workers had been kept busy studying and cataloging them. The old man was like a monk, carrying on a long tradition of study. I thought about d'Orbigny, now only decaying bones and dust in a cold grave—dust that was probably not unlike the fine patina that now covers the paper sheets and rocks of his legacy. But how could he be truly dead if human lives were in some small way still affected by his long-finished life? And, in the same way, the long-extinct species of life on earth are dead but not dead. The ammonites are gone, like countless other creatures, but their long presence on the face of the earth is felt still by the creatures that now live in the sea.

I finished my work in d'Orbigny's room and said my farewells to the museum staff. Just as I was leaving I heard peals of laughter coming from a nearby room, and peeked in. Two young paleontology students were tidying a desk covered with fossils, spiritedly discussing their plans for a Friday night to be spent in one of the world's most beautiful cities. They finally noticed me and looked at me curiously, for I was lost in reverie as I contemplated the clear skin and eyes of young men two decades my junior, neophytes beginning the grand adventure of science. I was the bridge between them and the man in d'Orbigny's office, a link between the young and the old, part of a chain extending backward and forward through time.

And so I adjourned to the park bench, to sit beneath the stone image of d'Orbigny, enjoying the sunshine but noting the first nip in the air, the unmistakable hint of fall amid the yellowing leaves in the garden. Finally I picked up the manuscript of this book. As I began to read, it occurred to me that the chapters dodge back and forth through time a bit like the hero in Kurt Vonnegut's *Slaughterhouse-Five;* time, in Billy's life, has lost its vector. Paleontologists live like that in some sense; their physical lives run along the simple, linear track of time, but their minds move back and forth through the ages, jumping onto the tracks where time moves at a more complicated pace. Perhaps we are searching for the lessons of our own survival, chasing clues by studying the lives of the Methuselahs.

Seattle, Biarritz, Paris
1989–1990

Acknowledgments

This book could not have been produced without the help of numerous people. I would especially like to thank Christine Hastings and my editor, Jerry Lyons at W. H. Freeman for their patience and advice. Sue Bolsson of the University of Washington proved once again that she is the greatest typist who has ever lived. Special thanks also go to my teachers, Drs. V. S. Mallory of the University of Washington and G. E. G. Westermann of McMaster University, for their support early in my career, and to my parents, son, and friends for support during various trying periods. Finally, I would like to thank the numerous paleontologists who have labored over the years to make sense of the fossil record. They are the heroes of this book.

INDEX